Mentoring and Diversity

Mentoring in Academia and Industry
Series Editor: J. Ellis Bell, University of Richmond, Virginia

---

Biology is evolving rapidly, with more and more discoveries arising from interaction with other disciplines such as chemistry, mathematics, and computer science. Undergraduate and Graduate biology education is having a hard time keeping up. To address this challenge, this bold and innovative series will assist science education programs at research universities, four-year colleges, and community colleges across the country and by enriching science teaching and mentoring of both students and faculty in academia and for industry representatives. The series aims to promote the progress of scientific research and education by providing guidelines for improving academic and career building skills for a broad audience of students, teachers, mentors, researchers, industry, and more.

Volume 1
Education Outreach and Public Engagement
by Erin L. Dolan

Volume 2
Active Assessment: Assessing Scientific Inquiry
by David I. Hanauer, Graham F. Hatfull, Deborah Jacobs-Sera

Volume 3
Getting the Most Out of Your Mentoring Relationships
by Donna J. Dean

Volume 4
Mentoring and Diversity: Tips for Students and Professionals for Developing and Maintaining a Diverse Scientific Community
By Thomas Landefeld

Thomas Landefeld

Mentoring and Diversity

Tips for Students and Professionals for Developing and Maintaining a Diverse Scientific Community

Thomas Landefeld
CSU Dominguez Hills
Carson, CA
USA
tlandefeld@csudh.edu

ISBN 978-1-4419-0777-6 e-ISBN 978-1-4419-0778-3
DOI 10.1007/978-1-4419-0778-3
Springer Dordrecht Heidelberg London New York

Library of Congress Control Number: 2009928621

Printed on acid-free paper

Springer is part of Springer Science+Business Media (www.springer.com)

Preface

Mentoring has always been an important factor in life and particularly in academia. In fact, making choices about educational pursuits and subsequent careers without input from mentors can prove disastrous. Fortunately, many individuals have "natural" mentors and for them these choices are greatly facilitated. Others are not privileged with natural mentors and as such often struggle with making these tough choices. Many times these individuals are from under served and disadvantaged backgrounds, where mentors are too few and far between. For them, deciding on which career path to take can be based not only on insufficient information but oft times on inaccurate information. Although the tips in this monograph are designed for helping all individuals who are interested in pursuing the study of science and science careers, a special mentoring focus is on those students who have not experienced the advantages of the privileged class. Additionally, tips are included for those who are interested in effectively mentoring these individuals. How and why a person gets to that point of wanting to mentor is not as important as the fact that they have made that commitment and this monograph will help them do exactly that.

When I received my PhD in Reproductive Endocrinology from the University of Wisconsin, I was ready and anxious to discover all kinds of new and exciting aspects about this field of science. Obviously, there was a lot that I needed to learn but I had the enthusiasm and drive to do that. Well, I proceeded to learn in two postdoctoral fellowships before going to the University of Michigan Medical School (UMMS) for my first faculty position. There, I did get my research program going, was funded by NIH R01 grants, and was indeed discovering some new and exciting aspects about the female reproductive system, using the sheep as an animal model. The reasons that I chose to obtain a PhD were now coming to fruition and I was enjoying my new career.

Along the way, I became involved with some administrative positions, first with the Medical School Admissions Committee and then as Assistant Dean for Research and Graduate Studies. I found that I enjoyed these roles and, in particular, an emphasis in working with the problem of under representation of minorities in the sciences. As a direct result of that interest and commitment, I then became Assistant Dean for Student and Minority Affairs as well. The more involved I became, the more I realized what a significant problem that this under representation of minorities in the sciences was. As such, my administrative tasks required more time away from my research, and it was then that I decided that I was going to focus my efforts on making a difference in this area of minority representation in the sciences. I subsequently left UMMS to assume a position at a Minority Serving Institution (MSI) and actually increased my efforts and commitments to this area, not only locally at my University but also nationally as part of several scientific societies and organizations.

In doing this, it became more and more obvious that one of the real contributing factors to the under representation of minorities in science, at all levels, was the lack of assistance that many of the students were getting, especially in the area of mentoring, an area that I had found in my experience working in this area to be invaluable. One way that I was able to address this was through Career Development Seminars and Short Courses at MSIs around the country supported by NIH, where I provided the students with information about careers as well as tips on how to be successful in achieving their goals within those careers, i.e., information that I gained from direct experiences throughout my faculty and administrative efforts over the years. Through these seminars, I made contact with thousands of students, many of which chose to continue to take advantage of this resource, even long after the visit has been completed.

This book is written to compliment the information that is provided to the students during these visits and, even more importantly, to provide it on a much larger scale than I can accomplish by traveling to individual schools.

I am especially proud of the commitment and dedication that I have made to this area but not nearly as proud of that as of the difference that my efforts have made with the many students in effectively addressing this continuing problem in science. I am hopeful that these efforts will ultimately result in us no longer using the words "under represented" when talking about minorities in science and that this book will help in accomplishing that.

Acknowledgments

A special thanks goes first to all of those individuals who have mentored me over my lifetime, both personally and professionally, especially Drs. Cloyzelle Jones, who took me through a part of my life where knowing how and when to deal with fighting the system made all the difference in the world in my successful struggle with equality in the academic world, Reverend Skip Wachsmann, who related to me on so many similar levels and, as such, was always there as for support, Derrick Bell, who not only served as a mentor for me but also as a role model for many others, and Dr. Frank Talamantes, the epitome of an advisor, role model, mentor, and friend. What I learned from all of my mentors, but especially those four, have allowed me to make a major difference in the lives of many, many others. I can only hope that others will feel the same way about the role that my mentoring has played in their lives.

I also want to thank all of those who not only supported me but also encouraged me as I "fought the system" for the betterment of many, despite the fact that their support often put them in situations where they were at risk. They, like me, subscribed to the Frederick Douglass' quote "Without struggle, there is no progress."

And, finally, and with no less importance, I want to than all those who I have mentored. Mentoring is a two-way street, i.e., without someone to mentor, mentors cease to exist! And, of course, the follow up is that those who have been mentored, in turn mentor others.

Contents

Chapter 1
Making Decisions About Careers in Science

The process of making decisions about a career in science represents a multifaceted one, beginning with what one actually wants to do as a professional, followed by what educational steps are necessary to "get there" and, of course, very importantly what one has to do to be not only competitive but also ultimately to be successful. None of these "steps" are accomplished easily and, more importantly, without input from others. In general, and specifically in academia, this input is referred to as advising and mentoring, with the latter really being the key component in the decision-making process. This mentoring comes from an assortment of individuals, very often with slightly different perspectives, which very importantly provides the mentee with an array of information pieces. As a result, a decision can be a well-informed one. However, despite this process being one that is commonly practiced, it is an "uneven" process in that some individuals have added advantages in making these decisions due to their exposure to mentors with strong backgrounds, experience, and status in the appropriate areas. Conversely, other individuals do not have some of these advantages and, as such, have been severely "undermentored" and therefore, underrepresented in the sciences. Identifying the reasons for this underrepresentation, and acting upon those reasons, is necessary to address and change these underrepresented numbers.

Unequal Advising and Mentoring

So what is tougher, deciding on what specific career that you want to pursue in the sciences or preparing sufficiently for that career once that decision is made? The truth is that both of these decisions are tough and quite honestly need input from many sources to be able to make the right decision. This is particularly true for students that have traditionally not had the advantages of the privileged group where fathers, mothers, friends, aunts, uncles, etc. have either had the exposure to these careers or actually are in them. For this reason, as well as others, women and individuals from underserved and often underprivileged groups, such as Blacks,

T. Landefeld, *Mentoring and Diversity,* Mentoring in Academia and Industry 4,
DOI 10.1007/978-1-4419-0778-3_1,

Hispanics, Native Americans, and Pacific Islanders, are grossly underrepresented in the sciences, at all levels, i.e., undergraduate, graduate/professional programs as well as in biomedical and health careers. Many efforts have addressed this problem and have been successful at least to some degree. For example, since financial support is always an issue in programs such as these, Federal agencies such as the National Institutes of Health (NIH) and the National Science Foundation (NSF) as well as a number of Foundations, e.g., Gates, Ford, Sloan, established programs designed to assist underrepresented students in their educational pursuits in the sciences, with the hope that this would in turn result in an increase in the numbers of these individuals choosing a career in the sciences. Although these programs, e.g., NIH Minority Access to Research Careers (MARC)/Minority Biomedical Research Support (MBRS) Programs, the NSF Louis Stokes Alliance for Minority Participation (AMP) Program, have been successful, the numbers are still disturbingly low. For example, when one examines the actual numbers of Blacks graduating with PhDs in the sciences or the number of Hispanic physicians, when compared to the growing proportion of these numbers comprising the general population, the numbers are rather dismal (Rodriguez 2008, "Still absent"; Wilson 2007, "Minority professors are underrepresented in top science programs, report says"; Nealy 2007, "Study: Minority faculty severely underrepresented in top STEM departments"; *Journal of Blacks in Higher Education* 2009a "The snail-like progress of Blacks into faculty ranks"; *The Journal of Blacks in Higher Education* 2009b "Notre Dame takes steps to increase Black faculty"). Moreover, when this fact is coupled with the current, national assault on these types of programs, e.g., Prop 209 (CA) and Prop 2 (MI), the future for changing these underrepresented numbers is not promising. In fact, this assault has resulted in a number of the programs that were designed to address the underrepresentation, and were in fact doing so, being eliminated, thereby making efforts to address the issue of underrepresentation extremely more difficult. Various attempts to address this in a "legal" way, e.g., refined admissions procedures (Journal of Blacks in Higher Education 2009c, "Will the new admissions procedures at University of California lead to an increase in Black enrollments?") as well as carefully assessing the legal wording of the laws (Hayes 2009, "Maryland AG offers legal guidelines for increasing diversity in State's universities") have been described and are in fact being assessed. With such efforts, academicians hope to still be able to strive for diversity relative to underrepresented minorities in the nation's colleges and universities so as to approach parity with the numbers of the ethnic groups comprising the general population. In fact, in the Hayes article, a number of justifications are listed for working for campus diversity. These include (a) the US is becoming more diverse daily; (b) citizens must become increasingly culturally competent in their communities and workplaces; (c) the nation must meet the needs of a multicultural public; (d) organizations demand culturally literate employees; (e) exposure to different cultures promotes improved understanding; (f) diversity promotes a robust exchange of ideas, especially in classrooms; (g) greater racial diversity is associated with enhanced critical thinking ability and (h) compositional diversity, interactional diversity, and diversity-related initiatives work interactively to improve educational outcomes. Moreover, the report cited in the article recommended that cultural diversity plans must be mission-driven and educationally focused. It is obvious that although these justifications do not include the need to specifically address the severe under-representation of individuals due to past discriminatory practices (for political reasons), this is inherent in any such efforts. In another report by Drs. Chubin and Malcolm (2008) (Inside Higher Ed, 2008, "Making a case for diversity in STEM fields"), they list issues that must be dealt with to achieve better representation in college STEM departments. These are (1) Presenting a clear articulation of the educational case for diversity, showing how students and society benefit, just as was reported in the Hayes article above; (2) the need for a more holistic way of thinking about diversity in STEM; and (3) the acknowledgement that stereotypes still matter and in turn they affect perceptions of quality and expectations for performance. The last issue, i.e., negative stereotypes, has been discussed for many years; in fact in the case of minorities and women it is better described as a "stereotype threat" and has been a major factor in the performance on standardized tests to the point when this "threat" is removed the students perform better (Moltz 2009 "The impact of negative stereotypes"). To take this even a step further, it was shown recently that in a 20-question test using questions from the verbal section of the graduate record examination (GRE), Black students performed better, i.e., the Black–white scoring gap disappeared after Obama's nomination and his subsequent election (*Journal of Blacks in Higher Education* 2009d "Can the 'Obama effect' help eliminate the Black–white scoring gap on standardized tests?"). If it indeed proves to be the case in larger studies, i.e., having a Black in the White House minimizes or possibly even eliminates the "stereotype threat," then efforts directed specifically toward this "threat" could really make a difference, especially as long as the use of standardized tests continue to be emphasized as a major determinant in admissions.

Unequal Representation

Also inherent within these justifications and issues is the need to address problems that are significant within our society and possibly none is bigger than the enormous problem of minority health disparities, which were in fact, recently referred to as "the most prevalent civil rights issue of the decade" (Nealy 2009a, "Racial health disparities called most prevalent civil rights issue of decade"). This article reports on the discussions that occurred at the 59th annual American Medical Student Association national convention. These health disparities that exist between minority and nonminority populations are a major concern, especially in the scientific community since these types of healthcare issues are exactly the reason that a number of individuals choose science as a career. This issue was in fact the topic of the article by Nealy (2009b) in *Diverse Issues in Higher Education* online (4/6/09) citing a UCLA study that outlined a strategy to bolster the number of Black HIV/AIDS researchers. As is so often the case with minority health disparities, Blacks are disproportionately affected by HIV/AIDS while the number of Black HIV/

AIDS researchers is small. Some of the strategies that were planned included the funding of research that encouraged partnerships between MSIs and minority organizations, such as the National Medical Association (NMA), loan repayment plans, cultural competency training, early exposure of Black college students to research and very importantly and relevant to this book, more Black HIV/AIDS mentors. A real "plus" in this particular area came directly from the Obama administration when it was announced that complacency about the nation's HIV/AIDS epidemic was going to undergo a direct attack through a media barrage (Sternberg 2009, USA Today "Putting AIDS back on the nation's radar"). It is planned that the Centers for Disease Control and Prevention will spend $45 million over 5 years for radio ads, transit signs, airport dioramas, online banner ads, and online videos, in both English and Spanish. The previous education campaign had been in 1987 and cost about $7 million. The new campaign will initially attach an urgency to knowing more about HIV/AIDS whereas later in the campaign, Blacks and Hispanics will be specifically targeted since they are disproportionately affected, e.g., Blacks account for 46% of all people with HIV, a tremendous "overrepresentation" when we compare that with the fact that Blacks make up only 12–13% of the population. It is hoped that this campaign will serve to complement studies like the UCLA one that is designed to get more Blacks involved in the research associated with HIV/AIDS. Regardless, studies like that UCLA one are certainly a step in the right direction; however, similar strategies have to be developed to also address not only the health disparities themselves, such as diabetes, prostate, and breast cancer, obesity, but also the low numbers of minorities involved in the research and treatment of these. At the same time, the research is very important; in fact it must be realized that many of the issues that contribute to these health disparities are not just science based, e.g., genetics, but also sociological, cultural, economical, and certainly political. For example, how much is obesity in the South Los Angeles community due to the lack of fresh fruits and vegetables at the local markets or the lack of fitness centers or the inability of residents to walk or run for exercise or the overabundance of fast food restaurants in the neighborhood? It is all of these factors, in addition to any possible "science" parameters such as metabolism, genetics, etc. that contribute to the health disparity problems. More will be discussed about this issue later.

In recent years, this problem of underrepresentation has become even more significant when assessing these numbers relative to gender, i.e., the number of underrepresented women earning degrees and continuing on into science careers have now greatly exceeded those for URM men. Moreover, this problem of the lack of Black and Hispanic men in education and science has now been identified as significant at all educational levels including high school graduation, where the 2005–2006 graduation rates nationwide for Black males was just 47% (Forde 2008a, "Report: Large school districts fail at graduating Black males"). For community colleges, which have traditionally been a major source of education for underrepresented minorities, the 2006 Integrated Postsecondary Education Data System survey (IPEDS) showed that Black males had a graduation rate of 16%, the lowest of all minority male community college students (Esters and Mosby 2007, "Disappearing acts; the vanishing Black male on community college campuses"). This has been identified as a problem even at Black colleges (Pope 2009a, b) in that less than one-third of men at Black colleges earn their degree in 6 years. More will be discussed regarding this problem later. As a result, this problem has developed to the point that there are now targeted programs specifically for Black and Latino males. Examples of these include the New York City College of Technology Black Male Initiative (BMI), the Student African American Brotherhood at Georgia Southwestern State University, and the Todd Anthony Bell National Resource Center on the African American Male at Ohio State University, which actually developed from the Governor's Ohio Commission on African American Males. Although these newly established programs have had to deal with those same antiaffirmative action foes, as mentioned earlier, the programs still represent a sincere effort to address a unique aspect of the underrepresentation of minorities in science problem (Moore 2008 "Recent report identifies factors that improve educational outcomes for Black males"; Ewers 2008, "Perspectives: Ohio Gov. focuses on graduating Black males, when will parents and mentors make similar commitment?"; Forde 2007, "BMI worth duplicating"; Esters and Mosby 2007, "Disappearing acts; the vanishing Black male on community college campuses"; Forde 2008b, "Black males in a state of emergency"; Heilig and Reddick 2008, "Perspectives: Black males in the educational pipeline" and Pluviose 2008, "Philadelphia mayor takes part in community college Black male summit"). In fact, in a recent article by Michelle Nealy (2009c) (2/11/09 *Diverse Issues in Higher Education*; "Black males achieving more on college campuses") examples of three individual success stories are given for Black males, due to the aforementioned programs. Although these models certainly represent ones to be duplicated, at the same time, "one size does not fit all." In fact, Dr. Shaun Harper is quoted in the article as saying "Borrowing a design from one campus and implementing it at another will not yield desirable outcomes for Black males. Trying to do the exact same kinds of things for Latino males would also not be effective." He goes on to say that "a better alternative is for institutions to establish a team of stockholders across the campus who are concerned about the disparity and empower them to come up with context-specific kinds of programs and interventions." Dr. Harper was speaking from a point of authoritative experience as his research has shown that 67.6% of Black male freshmen never complete their degrees. Also, his research has shown that only 4.3% of all enrolled students were Black males in 2003, the same percentage as in 1976! Moreover, although there are a number of reasons for this, he attributes much of the blame on the severe lack of engagement among Black students. This then goes directly to the point of how the successful programs, e.g., Ohio State, UGA accomplish better retention of Black males, i.e., both programs cite accessibility, connections, and engagement as the keys to success. Not surprising, both refer to programs at Morehouse College as models of success, which as one of the leading all-male campuses in the country has demonstrated outstanding success, graduating many leaders in many areas, over the years, including Dr. Martin Luther King,

Jr., Dr. David Satcher, Spike Lee, just to list a few. Certainly, engagement and connections represent critical components of mentoring.

Just as an additional comment on the possible effect of having a Black man as the US President, James Ewers (2009) in *Diverse: The Academy Speaks* (4/1/09) discusses the possible raising of the bar in his article entitled "African American boys can learn something from President Obama." Certainly the effect that role models can play on young people is well established and specifically the negative effect of the lack of positive role models for young Black men. This positive role model combined with other effects such as minimizing the stereotype threat, as mentioned above, can help to address this shortage of Black men in academia.

At the same time, even though the numbers for minority women who enter science programs and ultimately science careers have increased, there is still a major problem with these individuals not continuing on in those careers, especially in academia, a problem that is exacerbated by the changing demographics of the country and the resultant need for the face of the workforce to change. Also, issues such as minority health disparities, as mentioned above, will take on even more significance as the population changes, not only because there will be more individuals from those ethnic groups and therefore more disparities, but also because unless the underrepresentation in the medical sciences changes, those health professionals trained to treat the disparate diseases will not be, for the most part, from those groups that are most affected by these disparities. More on this issue of individuals staying "on course" will be discussed in Chapter 8 on professional careers.

Definitions of Minority and Diversity

As a note, at this time, it is important to mention that rather than using the word minority, for a number of reasons, both legitimate and not so legitimate, many institutions, whether they be academic, business, federal, or other, have chosen to use the word diversity, since that word does represent one that includes differences. Much of this change was due to the two landmark US Supreme Court decisions in June 2003 that defined the limits of affirmative action. In fact even more than just the use of the words diversity or multiculturalism instead of minority, a number of minority-targeted programs, i.e., race-exclusive eligibility, were modified and/or abandoned, primarily because of a concern about legal jeopardy, not because they were concerned that there was no longer a need to address underrepresentation (Schmidt 2006, "From minority to diversity"). Interestingly, with regard to the latter point, the rulings did not prohibit consideration of race outside the admissions context; however due to national, legal pressure from not only the groups, e.g., Center for Equal Opportunity and the American Civil Rights Institute, but also with the support and assistance of the Office for Civil Rights and the U.S. Justice Department, many schools chose to abandon or modify their programs, often resulting in a change in the goal of addressing the underrepresentation issue. In fact even those programs that were modified to no longer be "minority-exclusive" still resulted in less of a commitment to underrepresented groups, thereby resulting in a decrease in the number of underrepresented students being served. Thus, for many individuals who had advocated and expended considerable efforts in truly addressing the underrepresentation, those efforts effectively set back the strides that had been made in going forward to address the problem. However, it is important to define the types of differences that the organizations are addressing with the word diversity because it will not always be the underrepresented, underserved populations. For example, diversity of thought, of ideas, of disciplines are all important to an organization and, in fact, very often that can indeed be effected by diversifying the ethnic populations involved. However, especially in the scientific community, without the inclusion of those individuals from ethnic backgrounds who have heretofore been excluded, or only included minimally, the future of not only the scientific community but society, in general, is at risk (Jaschik 2008, "A new look at the impact of diversity"). For example, is a program comprised of white men with different color hair a model for diversity? Most would agree that it is not unless one is perhaps referring to a hair salon or cosmetology classroom! In education, and particularly the science community, diversity has to refer to ethnic minorities due to the dearth of individuals from certain ethnicities that are participatory in that scientific community. Also there is the question of what role the faculty play in diversity since in the past this has not been a focus of many institutions and therefore not of their faculty members (Jaschik 2009 "Doing diversity in higher education"). First and foremost, faculty are in a unique position when compared to diversity and student affairs officers in that they see first hand the benefits of diversity in the classrooms through discussions, student enrichment, and a general sense of the climate, i.e., welcoming or hostile. Second, depending upon what department they are in, their efforts could be frustrating and challenging, i.e., in traditional STEM departments it is most gratifying when they actually see the types of changes that are necessary. On the other hand, if they are in an ethnic studies department, they face a different type of challenge in that the students may be more diverse but the institutional very often does not institutionalize them as readily, due to these being part of an "ethnic studies" program already. Beyond these two points, it is then critical for the institution to recognize, acknowledge, and reward the efforts made by faculty toward diversity. Most of the individuals who make these types of commitments are certainly not doing it to be rewarded; however at the same time, the system is set up with this type of "reward" structure and when efforts are not rewarded, it sends a message to everyone involved that it is not important. For this reason, a commitment to diversity should definitely be a criterion for faculty hiring but only if the institution has chosen to recognize diversity as part of its mission. To this point, *The Journal of Blacks in Higher Education* (2009e) reported that the results of "The American College Teacher: National norm for the 2007-08HERI Faculty survey" showed that more than 75% of college faculty say that they work to "enhance students'" knowledge of and appreciation for other racial/ethnic groups." This represents an increase of 17.6% points over a survey taken 3 years earlier. Although there are many benefits to committing to diversity and what occurs as a result of those efforts, perhaps the most important one is in the area of mentoring, the topic of this book. Whether it is mentoring others, who then go onto mentor others or receiving mentoring as I did specifically when I chose to join the battle for equality, the rewards and gratification from this are immeasurable. This is almost universally recognized but is not typically acknowledged in the academy in the same way that outstanding research or excellent teaching is. In fact, the inclusion of the commitment/activities to diversity that Virginia Tech placed as part of tenure and promotion decisions has been attacked by the Foundation for Individual Rights in Education (FIRE) to the point that a letter from the organization stated that the language on diversity amounts to promotion of a "political orthodoxy" thereby representing an ideological loyalty oath (Wilson 2009, "Critics challenge diversity language in Virginia Tech's tenure policy"). In addition, in an article on the National Association of Scholars website, it is stated that a commitment to diversity does not belong in guidelines that evaluate a professor's bid for promotion and tenure and that "diversity" is not a category of academic accomplishment equivalent to high-quality teaching or success in scholarly research and publishing. The article in fact says that "diversity is an ideology." On the other hand, a recent article by Dodson (2009) in *Diverse Issues in Higher Education online* (3/27/09) discusses a recently published book entitled *Doing Diversity in Higher Education: Faculty Leaders Share Challenges and Strategies* edited by Winnifred R. Brown-Glaude. The book actually examines the strategies of faculty to serve as "change agents" encouraging diversity on campus based on case studies from 12 universities, including Columbia University, Spelman and Smith Colleges as well as some state universities. Moreover, Dr. Brown-Glaude also uses her experiences as a professor of sociology at Bowdoin College where she introduced students in her class to "multiple perspectives" on the subject of diversity as well as her experience as director of a 4-year study, "Reaffirming Action: Designs for Diversity in Higher Education." The arguments put forth in the article and book are very similar to those that have been cited above, including, as stated in the article, the fact that voices of educators and scholars who advocate diversity and espouse its benefits for education "are rarely heard in public." She states that as a result of this the debate around diversity is often hijacked by the opponents who keep the focus on race, such as quotas and therefore racial preferences. The article goes on to state that even the advocates have a tendency to focus on the legacy of discrimination, rather than on the educational benefits of having a more representative, inclusive college population. The recommendation is that there are more "bottom-up" actions; however, as mentioned already and confirmed in this article, "diversity work" is not valued or rewarded, especially when faculty members seek tenure. As a result, most of that type of work is done by senior tenured educators who are secure in their positions. However, even with those individuals, there is often a reluctance to act and become involved in these types of issues due to the risks of "challenging the system," e.g., how will it affect my standing in the department, in the college, nationally? My experiences confirm and validate this type of fear. When I was at the University of Michigan Medical School, as a tenured Associate Professor, but advocating and speaking out about changes needed in the commitment to minority programs, I was eventually "constructively demoted," which in academic terms means that although I was not fired nor was my title removed, many other actions were taken such as removal from committees and other activities that related to minority affairs, reduction in laboratory space, no raises, etc. As a result, one still has a "job" but without most of those "things" that mean the most to what one does In that sense, tenure may mean "lifetime security" but is really not a protection of academic freedom. Dr. Brown-Glaude also discusses the concern of President Obama's election in terms of these programs, in that many opponents will say that diversity is no longer an issue now that there is a Black President. Her answer to that is that we must have conversations that address those concerns; conversations that much of America and academia have not wanted to have in the past.

Thus, this book is written to provide assistance for all students desiring a career in "science" and, in particular, to those students who are not routinely made aware of this information as they navigate their way through the academic maze, which is made more difficult due to their ethnicity and cultural backgrounds. The emphasis will be on mentoring and how this process can provide students with guidelines, directions, advice, and general suggestions that will facilitate their choices about career more easily and strengthen their preparation for the programs that are necessary for them to complete before they can enter into these careers. However, and very importantly, the contents within this book will also assist those individuals in their chosen careers to not only maintain themselves in their professions but also, most importantly, succeed in those careers. Finally, and just as importantly, the book will address issues associated with mentoring from both the student and faculty perspective, as many faculty members often find difficulty in mentoring students due to a lack of understanding, sensitivity, and familiarization.

References

Chubin DE, Malcolm SM (2008) Making a case for diversity in STEM fields. Inside Higher Ed http://www.insidehighered.com/views/2008/10/06

Dodson AP (2009) Broadening the definition of diversity. Diverse Issues in Higher Education online http://www.diverseeducation.com/artman/publish/2009/03/27

Esters LI, Mosby DC (2007) Disappearing acts; the vanishing Black male on community college campuses. Diverse Issues in Higher Education online http://www.diverseeducation.com/artman/publish2007/08/23

Ewers J (2008) Perspectives: Ohio Governor focuses on graduating Black males, when will parents and mentors make similar commitment? Diverse Issues in Higher Education online http://www.diverseeducation,com/artman/publish/2008/07/15

Ewers J (2009) African American boys can learn something from President Obama. Diverse issues in Higher Education The Academy Speaks http://www.diverseeducation.com/artman/publish/2009/04/01

Forde D (2007) Black male initiative worth duplicating. Diverse Issues in Higher Education online http://www.diverseeducation.com/artman/publish/2007/11/29

Forde D (2008a) Report: Large school districts fail at graduating Black males. Diverse Issues in Higher Education online http://www.diverseeducation.com/artman/publish/2008/07/30

Forde D (2008b) Black males in a state of emergency. Diverse issues in Higher Education online http://www.diverseeducation.com/artman/publish/2008/06/10

Hayes D (2009) Maryland AG offers legal guidelines for increasing diversity in state's universities. Diverse Issues in Higher Education online http://www.diverseeducation.com/artman/publish/2009/03/16

Heilig JV, Reddick RJ (2008) Perspectives: Black males in the educational pipeline. Diverse issues in Higher Education online http://www.diverseeducation.com/artman/publish/2008/08/13

Jaschik S (2008) A new look at the impact of diversity. Inside Higher Education 2006 Integrated Postsecondary Education Data System survey (IPEDS) http://www.insidehighered.com/news/2008/12/19/diversity

Jaschik S (2009) Doing diversity in higher education. Inside Higher Education http://insidehighered.com/2009/01/12/diversity

Journal of Blacks in Higher Education (2009a) The snail-like progress of Blacks into faculty ranks http://www.jbhe.com/latestnews/1-1-09

Journal of Blacks in Higher Education (2009b) Notre Dame takes steps to increase Black faculty http://www.jbhe.com/latest/index Accessed 3/26/09

Journal of Blacks in Higher Education (2009c) Will the new admissions procedures at the University of California lead to an increase in Black enrollment? http://www.jbhe.com/latest Accessed 3/5/09

Journal of Blacks in Higher Education (2009d) Can the "Obama effect" help eliminate the Black-white scoring gap on standardized tests? http://www.jbhe.com/latest/news/2-12-09

Journal of Blacks in Higher Education (2009e) Survey finds that college faculty are more likely to value racial diversity programs than was the case three years ago http://www.jbhe.com/latest/index Accessed 4/2/09

Moltz D (2009) The impact of negative stereotypes. Inside Higher Ed http://www.insdiehighered.com/news/2009/02/25

Moore J (2008) Recent report identifies factors that improve educational outcomes for Black males. Diverse Issues in Higher Education online http://www.diverseeducation.com/artman/publish/2007/08/23

Nealy MJ (2007) Study: Minority faculty severely under represented in top STEM departments. Diverse Issues in Higher Education online http://www.diverseeducation.com/artman/publish/2007/10/31

Nealy MJ (2009a) Racial health disparities called most prevalent civil rights issue of decade. Diverse issues in Higher Education online http://www.diverseeducation.com/artman/publish/2009/03/16

Nealy MJ (2009b) UCLA study outlines strategy to bolster number of Black HIV/AIDS researchers Diverse issues in Higher Education online http://www.diverseeducation.com/artman/publish/2009/04/06

Nealy MJ (2009c) Black males achieving more on college campuses. Diverse Issues in Higher Education online http://www.diverseeducation.com/artman/publish/2009/02/11

Pluviose D (2008) Philadelphia mayor takes part in community college Black male summit. Diverse issues in Higher education online http://www.diverseeducation.com/artman/publish/2009/04/09

Pope J (2009a) Under a third of men at Black colleges earn degree in 6 years. USA Today 3/30/09

Pope J (2009b) Men struggling to finish at Black colleges. Diverse issues in Higher Education online http://www.diverseeducation.com/artman/publish/2009/03/30

Rodriguez E (2008) Still absent. Progress Magazine, vol 5 No.!

Schmidt P (2006) From minority to diversity. The Chronicle of Higher Education http://chronicle.com/weekly/v52/122/22

Sternberg S (2009) Putting AIDS back on the nation's radar. USA Today 04/08/09

Wilson R (2007) Minority professors are under represented in top science programs, report says. The Chronicle of Higher Education http://chronicle.com/daily/2007/11/565n Accessed 11/1/07

Wilson R (2009) Critics challenge diversity language in Virginia Tech's tenure policy. The Chronicle of Higher Education http://chronicle.com/daily/2009/03/14550n Accessed 3/26/09

Chapter 2
Mentors and Mentoring

Although there are many definitions of mentors and mentoring, there are certain common themes across the many definitions. For example, someone who coaches and tutors is often called a mentor. Someone who advises and assists in both the individual's personal and professional life is often considered a mentor. There are cases where an organization and/or an institution serve in a mentoring capacity. The important point is that in all cases, a mentor makes an individualized, personalized effort to assist someone in achieving their goals, reaching their objectives, and/or becoming successful. And, although there are often common themes, it is the individualization of the efforts that makes mentoring really work, especially, as mentioned above, for the disadvantaged individual, regardless of how that disadvantage may be exhibited, e.g., educationally, socioeconomically, etc.

Definitions of Mentors and Mentoring

So what is a mentor (and the process of mentoring) and why do we hear so much about this, not only from professors and other academicians, but also from athletes, celebrities, organizations, companies, and all kinds of other groups?

Let us start with the definition. Mentor, as defined by Random House Unabridged Dictionary, is (1) a wise and trusted counselor or teacher, (2) an influential senior sponsor or supporter, (3) to act as a mentor (v), and (4) to act as a mentor to (v). Synonyms are listed as advisor, master, guide, and preceptor. So, first it is important to recognize that the word "mentor" can be used as a noun to describe a trusted advisor/preceptor and also as a verb to describe a process, i.e., to mentor or guide someone. Both definitions are critically important in the context of this series of texts as well as for this specific book in that mentors need to be both identified (noun) and used (verb). One of these does not work without the other although there is a temporal sequence involved, i.e., one cannot be mentored without mentors! In most cases, this represents a one-on-one relationship although these relationships can become part of a larger mentoring program, designed to provide opportunities to students, employees,

T. Landefeld, *Mentoring and Diversity,* Mentoring in Academia and Industry 4,
DOI 10.1007/978-1-4419-0778-3_2, © Springer Science+Business Media, LLC 2009

and others. For example, there are a number of organizations that are built on the concept of mentoring such as http://www.mentoring.org and http://www.mentoring-group.com. Similarly, companies such as Genentech, Intel, and Human Genome Sciences all have mentoring programs in place to assist them in improving various aspects of their organization through mentoring initiatives (Alberty 2009, "Mentoring to the bottom line"). In fact, two companies, Propel and Bioadvance, represent mentoring programs for early-stage life-science entrepreneurs and, as such, address a need for mentoring in the biotech industry (Grant 2009, "The mentorship market"). Moreover, as will be mentioned, and in fact, elaborated on later, many of the Federally funded programs designed to address the underrepresentation of minorities in the sciences utilize mentoring paradigms to be successful. So how do we define mentoring? The Web site http://www.mentoring.org defines it as "a structured and trusting relationship that brings young people together with caring individuals who offer guidance, support and encouragement aimed at developing the competence and character of the mentee." The Office of Research, Consumer Guide (http://www.ed.gov/pubs/OR/ConsumerGuides/mentor) defines mentoring "…from the Greek word meaning enduring, as a sustained relationship between a youth and an adult." They go on to state that "through continued involvement, the adult offers support, guidance, and assistance as the younger person goes through a difficult period, faces new challenges or works to correct earlier problems." They also add that "In particular, where parents are either unavailable or unable to provide responsible guidance for their children, mentors can play a critical role." There are many other definitions but most all of them will include the words guidance, assistance, and support. Other words and phrases that are commonly used to describe mentors include a good listener, caring, compassionate, committed, and dedicated. Despite the commonalities associated with the definitions, some might modify the definition in that it is not always an older person to younger person relationship (even though that is more typical) since anyone can be mentored at any stage of their life, especially as they embark on a new path or go in a new direction. As a personal example, I was mentored scientifically by a number of individuals in the scientific research arena throughout my education as a biomedical scientist. These mentors were professors, both in and outside of the classroom, researchers, both in and outside of the laboratory, and academicians in all parts of the academy. However, when I became intricately involved in the fight for equal rights within academia, it was a senior educator who had been directly involved in the 1960s Civil Rights demonstrations that truly mentored me as to how to fight this battle, teaching me things that I had not experienced, realized or been taught by previous mentors. Thus, I, like many, would profess that mentors, and what they do, are an essential part of everyone's life with times and situations where these take on even more importance and significance. One of these more critical times (and situations) is in the field of education and the subsequent decisions regarding educational pathways and career choices. For that reason, mentoring should especially be a part of a young person's life as early in their education as possible. And, as one would imagine there is what is referred to as "natural" mentoring, i.e., that which is provided by parents, teachers, ministers, and coaches during their childhood. Unfortunately, not all children are afforded the same exposure and opportunities relative to these mentors for a number of reasons. As such, their decisions are often not influenced positively or, even worse, influenced negatively, and not just related to education but life in general. This is the reason that there are many mentoring organizations and programs that represent what is referred to as "planned" mentoring. For example, these organizations often target "at risk" youth or youth who do not have access to those natural mentors, either at all or as readily and as such do not often have the "natural mentors." Examples of such organizations include Big Brothers/Big Sisters of America, Help One Student to Succeed, One Hundred Black Men, and The National One-on-One Mentoring Partnership. Regardless of whether the individual comes from an advantaged background or a disadvantaged one, many young people have benefitted from such organizations as evidenced by personal testimonials provided by the organizations.

A critical aspect of mentoring and, in particular, the planned mentoring programs is matching the mentor and mentee (some prefer to use the word protégé). This match is truly the heart and soul of the mentoring process. In some natural mentoring scenarios, there is no reason for a "match," e.g., a parent and a child, a teacher and a pupil, a coach and a player; however, still, even in these cases, especially the latter two, the match may not be optimal and therefore, not effective, i.e., the mentor has a title, or is in a position, to mentor but may not choose to do so, at least in an effective manner. In those cases both the mentor and the mentee hopefully recognize this and make the change. This does not necessarily take away from a teacher's effectiveness as a teacher, or the coach as a coach, but only as a mentor. In those cases of "planned mentorships," the match can actually be decided upon through meetings, interviews, surveys, and/or other interactions. Regardless, the matching is critical as it can move the "planned mentoring," in many cases, into almost a "natural-mentoring" situation.

There has actually been some debate as to whether or not a mentor/mentee relationship should be a "friendship" (Compact for Faculty Diversity's Institute on Teaching and Mentoring Lederman 2008 "Mentor, friend – or both?"). Although in some ways, the debate was really a matter of semantics or rhetoric, it did raise some interesting points. Comments included: "There must be an emotional connection and a level of caring but that 'friendship' was not the correct word."; "liking is not necessarily a part of the relationship" and "personal issues are outside of what was part of the relationship." The bottom line is that there are always limitations in relationships and the mentor–mentee situation is no different. Moreover, those limitations will vary among relationships depending upon the individuals involved, especially since mentoring is totally about personalization and individualization. Regardless of what the final analysis is relative to friendship as part of mentor relationships, the interactions between the mentor and mentee have to be involved and emotional (Evans 2008, "Mentoring magic: How to be an effective mentor; tips from two highly successful principal investigators"; National Academy of Sciences 1997, "Adviser, Teacher, Role Model, Friend – on being a mentor to students in science and engineering"; Hayes 2008, "Community college mentoring programs help students stay engaged"; Hannah 2007, "How to build a successful mentorship"; Cole 2007, "How to be coachable and get the most from your mentor"; National Science Foundation 1997, "Mentoring for the 21st Century"; http://www.NSDC.org).

Mentoring Paradigms

As one continues to consider decisions regarding education, even for those young people who make some "good" decisions and successfully make it into some aspect of the educational system, with or without natural/planned mentoring, they are still faced with decisions that continue to require input and assistance from mentors (most likely new mentors with different experiences that now can assist them with the next stage of their life). In fact, one should never depend solely on past mentors but should continue to identify new mentors throughout life, most often as one moves into a new life phase, e.g., the case of me being mentored at a stage of my career when I moved into a new arena of commitment. As another example, a mentor who is a high school teacher most likely will become limited in academically mentoring a student as the student makes plans to attend a professional school, e.g., medical school, even if the teacher is a science teacher. Similarly, although a counselor in an undergraduate careers office can effectively mentor a student throughout their undergraduate experience, when the student requires assistance in a very specific discipline, mentoring from a "generalist" in a Careers Office becomes more difficult. From this, it should be obvious that people need to have a number of different mentors throughout their lifetime as it is unlikely that one mentor can serve all purposes and needs, especially as situations and circumstances change throughout life. At the same time, any mentor who really knows the mentee well can always provide assistance and support just based on their intimate knowledge of the person. This is what is usually referred to as personal development mentoring. A key component of any successful mentoring relationship is the recognition that it is a two-way street, i.e., a commitment must be made by both the mentee and the mentor. An excellent example of this was recently published by http://www.aamc.org/gradcompact entitled "A Compact between biomedical graduate students and their research advisors" (12/08). In that publication, not only are points made regarding the commitment that must be made at all levels, i.e., faculty member, student, program/department, and institution, but also a delineation (listing) of the commitments by the student (mentee) and faculty member (mentor) is almost a contractual design. A signed contract is not necessary but a demonstration of the commitment on the part of both is essential for the mentorship to be successful. Interestingly, but not surprising, this Compact was modeled on the AAMC Compact Between Postdoctoral Appointees and Their Mentors which was developed by the same AAMC Group on Graduate Research, Education and Training (GREAT). It is not surprising since mentorship, as indicated above, requires very similar actions, commitments, and involvements regardless of the stage of one's career. As such, a compact for the mentoring of young faculty members would look very similar. The key to the success of any of these, of course, is that the compact is honored by all those associated with it. This will be referenced again in the chapter on making choices about graduate school vs. other health professional schools.

It should be noted that person-to-person, one-on-one mentoring works best as the mentoring process is really all about personalization and individualization. We all certainly know that facial expressions, body language, and other gestures are important in communication and, despite all of the modern communicative technologies, these iterations are essentially missing from all forms of communication except for personal interactions. However, at the same time, the use of modern communications, e.g., email, texting, etc., can certainly bring a person who is in one part of the country essentially "next door" to another. This is not necessarily ideal but it serves a very meaningful purpose, especially when it is not possible to meet in person and often as a follow up, i.e., once the initial meeting and interaction has occurred. For example, when I give a lecture, e.g., career development, at a school in another part of the country from where I live and work, I talk with students regarding career choices. Since I am only there for a day or two, our discussions are limited. However, in many cases, the students stay in touch regarding further questions, editing of personal statements, information for completing applications, etc. In other words, I mentor these students through email correspondence, and in some cases, will meet up with them again at national conferences and/or during another visit to the campus to further mentor them. The addition of this "long-distance" mentoring has provided an important component to the process of mentoring in that a time continuum can be effectively established. Moreover, maintaining that type of contact is especially critical to the mentoring process. One national organization, i.e., MentorNet (http://www.mentornet.net) has had great success with e-mentoring in addressing the underrepresentation of women in science and engineering (Gawrylewski 2008, "Taking Mentorship Online").

Although much of the evidence for mentoring success is anecdotal through testimonies and other recounted stories, mentoring unquestionably does work. Simply ask anyone who mentors on a regular basis and they will recount one success after another. They can do this because the mentoring relationship continues much beyond the initial interactions; in fact it usually persists through much of a lifetime. Similarly, ask anyone about their mentors and they can tell you of their experiences and how those experiences made a major difference in their lives. For these reasons, almost everyone who has been mentored at some time during their life will mentor someone else, very often using the knowledge about mentoring that was learned when they received mentoring.

There has been an emphasis in recent years on Learning Centers and in particular, the Science of Learning Centers (SLC). In fact, in 2003 the National Science Foundation (NSF) established the Science of Learning Centers (SLC) Program. This program was developed to enable large-scale, comprehensive, and scientific study of learning. Despite this science of learning being a multidisciplinary, multilevel program, sometimes even a "center," at the same time, a critical part of it is the individuality component, i.e., mentoring since mentoring always represents a rich educational and training opportunity. An additional advantage, of course, is that if the Center is financially supported, then support for the mentoring efforts can often be in place.

Benefits of Mentoring

A very important aspect of mentoring, especially right now in academia, is mentoring young faculty members as many senior faculty members are at, or near, retirement. Plus, unfortunately, faculty mentoring has never been a high priority at most institutions, possibly because it is not easily taught and also since often it is "just assumed" that the faculty member knows how to mentor. For all of these reasons and more, programs are being developed to mentor young faculty. Some examples are given in a report by Audrey Williams June (2008) (*The Chronicle of Higher Education* "A helping hand for young faculty members") and include both formal and ad hoc programs. In the minority scientist arena, this problem is even more exacerbated by the fewer number of faculty already as discussed in the article entitled "Who will fill their shoes" by Nealy (2008) (*Diverse Issues in Higher Education online* 11/13/08).

One additional aspect of mentoring can be considered that of advocacy, since the more that the mentor knows about the mentee, the more support that can usually be provided. Often, this is in the form of a letter and/or verbal recommendation. However, it can also rise to another level of advocacy, depending upon the situation. For example, as the Assistant Dean for Student and Minority Affairs, one of my responsibilities was to advocate for the students in a variety of situations. One of the first of those was when four minority students were being recommended for dismissal from school. Although I could have advocated for them, simply as part of my job, that would not have been nearly as effective as advocating for them as students who I had mentored and therefore knew well, including the circumstances surrounding the recommendation for their dismissal. As it turns out three of the students were allowed to remain in the program and went on to receive their medical degrees. Although there were a number of factors that had to be considered in the school ultimately making such a decision, I know that the fact that I had mentored these students made a difference in my advocating for them. In other examples of advocacy, mentors provide letters of recommendation, calls to admissions committee members as well as a number of other cases, all of which are most meaningful and beneficial because of the mentoring relationship that they have with their mentees.

What does mentoring really do? Academically, it promotes achievement which in turn improves retention (and decreases attrition). Career wise, it assists mentees in setting goals, in advancing, in networking, and in making changes, if so desired. A recent study even related the lack of appropriate mentors and mentoring as a contributory factor in those scientists who have committed fraud in their research studies, which is not totally surprising since mentors play a significant role in teaching ethical behavior as part of conducting good research (Brainard 2008, "Scientists who cheated had mentors who failed to supervise them"). In an article entitled "Carnegie Foundation creates new 'owner's manual' for doctoral programs" (Wasley 2007), it is stated that the study recommends that doctoral programs adopt new structures that allow students to have several intellectual mentors and come to think of mentorship as less an accident of interpersonal chemistry and as more a set of techniques that can be learned, assessed, and rewarded. Models of this are given as Arizona State University that awards an annual $5,000 prize to an "outstanding doctoral mentor" and University of California that places new graduate students in "mentoring triplets" with both a faculty mentor and a more experienced graduate student. On a personal basis, mentoring improves the mentee's self-esteem, supports changes and adjustments, improves their ability to interact and communicate, and provides them with someone who they can truly lean on. If one considers these few examples alone, it is immediately obvious why mentoring of minorities and women is essential, since in most areas of academia, attrition rates are almost always higher for women and minorities.

References

Alberty B (2009) Mentoring to the bottom line. The Scientist 20(9):79

Brainard J (2008) Scientists who cheated had mentors who failed to supervise them. The Chronicle of Higher Education http://chronicle.com/daily/2008/08/4405n Accessed 8/29/08

Cole Y (2007) How to be coachable and get the most from your mentor. DiversityInc http://www.diversityinc.com

Lederman D (2008) Mentor, Friend-or Both. Inside Higher Ed http://www.insidehighered.com/news/2008/10/28/mentor

Evans J (2008) Mentoring magic: How to be an effective mentor; tips from two highly successful principal investigators. The Scientist 22(12):70

Gawrylewski A (2008) Taking mentorship online. The Scientist 21(7):83

Grant B (2009) The mentorship market. The Scientist 23(1):58

Hannah DC (2007) How to build a successful mentorship. DiversityInc http://www.diversityinc.com/public/3646

Nealy MJ (2008) Who will fill their shoes. Diverse Issues in Higher Education online http://www.diverseeducation.com/artman/publish/2008/11/13

Hayes D (2008) Community college mentoring programs help students stay engaged. Diverse Issues in Higher Education online http://www.diverseeducation/artman/publish/2008/07/24

June AW (2008) A helping hand for young faculty members. The Chronicle of Higher Education http://www.chronicle.com/weekly/v55/i03/03a010019

National Academy of Sciences, National Academy of Engineering and Institute of Medicine (1997) Adviser, Teacher, Role Model, Friend – on being a mentor to students in science and engineering, National Academy Press

National Science Foundation (1997) Mentoring for the 21st Century

Wasley P (2007) Carnegie foundation creates new "owner's manual" for doctoral programs. The Chronicle of Higher Education http://chronicle.com/daily/2007/12/868n Accessed12/4/07

http://www.aamc.org/gradcompact Compact between Biomedical Graduate Students and Their Research Advisors

http://www.mentoring.org What is mentoring?

http://www.NSDC.org Mentoring the mentor: A challenge for staff development

http://www.mentoringgroup.com The infrastructure of effective mentoring

http://www.ed.gov/pubs/OR/ConsumerGuides/mentor Mentoring

Chapter 3
Minorities in Science

The underrepresentation of ethnic minorities in science has been an issue for many decades and has been exhibited at all levels of the educational pathway, including of course professional science careers, which is not surprising since these careers require the successful completion of educational levels. Thus, if the underrepresentation is seen at K-12, college, and professional/graduate school levels, then underrepresentation has to be exhibited at the career level. Generally this has been referred to as the educational/career pipeline and the absence of adequate numbers along the way referred to as "leaks in the pipeline." For ethnic minorities, a number of reasons for these leaks have been identified and actually some remedies have been identified as well. Unfortunately this issue has not been enough of a priority for those solutions to have been implemented, very often due to inadequate amount of funding, so the problem has continued. Now, considering the changing demographics of the country, this problem is becoming an even major issue with regard to the scientific workforce and the resultant advancement of science in the future.

Underrepresentation Numbers

The number of individuals from certain ethnicities has long been underrepresented in the sciences, i.e., the percentage of these individuals in the area of science is significantly lower than their percentage in the general population. These groups include Blacks, Native Americans, Hispanics, and Pacific Islanders. Thus, one must recognize that there is a distinction between a minority, which conventionally has been nonwhites due to the white population being the majority in USA, and an underrepresented minority, i.e., URM. As indicated, those groups mentioned above have long been underrepresented at all levels of science, i.e., undergraduate majors, graduates, graduate and health professional programs as well as in the professions, e.g., physicians, dentists, pharmacists, science researchers, the science professoriate, etc. These groups are not only underrepresented but also grossly underrepresented. For example, in 2005, despite Hispanics comprising 14% of the

T. Landefeld, *Mentoring and Diversity,* Mentoring in Academia and Industry 4,
DOI 10.1007/978-1-4419-0778-3_3, © Springer Science+Business Media, LLC 2009

nation's population, just 7.5% of the bachelor's degrees in engineering were earned by Hispanics. Similarly in 2005, Hispanics earned just 7.5% of the bachelor's degrees in biology, 6.8% in computer sciences, 6.5% in physics, and 5.8% in mathematics. Moreover, considering that Blacks, Hispanics, and Native Americans comprise currently about 33% of the general population, only about 6% of the physician work force is represented by individuals from these ethnic groups. Another example is in the area of doctoral degrees awarded in the sciences, both life and physical. Of the 9,329 PhDs awarded in 2006 in the fields of science, only 6% were to Blacks, 4.2% to Hispanics, and 0.2% to Native Americans, for a total of 10.4%, again considering that these groups currently comprise more than 30% of the population. Even when one considers total PhDs and not just those in the sciences (Lederman 2007, "A Haven for Minority scholars"), of the approximately 55,000 PhDs awarded annually, only about 5,000 total are awarded to Blacks, Hispanics, and Native Americans (~9% when the total percentage of the population represented by those groups is almost 30% and rapidly growing). In terms of the "pipeline leak" for underrepresented minority students from K-12 through STEM PhD recipients, the data is staggering [CPST, data derived from National Science Foundation (1997) (NSF), WebCASPAR; database, National Center for Education Statistics, Digest of Education Statistics, 2006 and US Census Bureau, Population Division], as the URM numbers go from 35.5 to 5.6%. On the other hand, non-URMs only go from 63.2 to 49.9%! Perhaps even more of a concern is that the numbers for the group identified as Non-US citizens and other/unknown race/ethnicity go from 1.3 to 44.5%, with the number of international applications to graduate school continuing to increase (Redden 2009, "International applications up 4%"; Schmidt 2009, "Doctoral universities pull ahead in competition for foreign graduate students"). Moreover, based on the survey by the Council of Graduate Schools (CGS) where almost 250 institutions responded, not only are these applications rising but also they are showing the largest increases at the doctoral institutions where the largest number of international students is already enrolled. In another example of a focus on international students, David Moltz (2009a) reports in an *Inside Higher Ed* article (04/08/09) titled "Diversifying Middle America" that officials from the Northern Wyoming Community College district reported on their efforts to diversify their campus. First, it is noteworthy to recognize that nearly 97% of the students who attend the two colleges in that District are white with Native Americans from traditionally local tribes making up the largest minority group and that many of these minorities are there because of its National Junior College Athletic Association sports teams. The bottom line regarding their diversification efforts is that they are focused on attracting international students using their strong transfer history, their inexpensive costs, and their self-branding as "the gateway to the American West" as selling points. Is this "diversifying middle America"? Certainly not in terms of dealing with the underrepresented minority groups in science which are the focus of this book. Interestingly, even with the presence of these international students in Wyoming, incidents have occurred, e.g., a "dark" international student has been stopped five times by police with no citations given! It is obvious that the leaks seen for URMs from K-12 to PhDs in STEM are in stark contrast to these international numbers and, as such, are significant since, at the same time, the numbers for non-US Citizens are increasing significantly. As for the numbers of underrepresented minorities in the faculty ranks, from the Nelson (2007) "Diversity Surveys 2007", it was reported at the 2008 CGS meeting by Dr. Daryl Chubin that the numbers of tenured/tenure-track faculty in Top 50 departments, e.g., biology, psychology, chemistry, and mathematics for Black, Hispanic, and Native Americans did not exceed 3.6% (i.e., Hispanics in psychology) and in fact were all in the 1–3% range, whereas for whites, the percentages were all above 80%. In addition, very often differences are seen within the various underrepresented groups, in this case among students graduating, as evidenced by a report by Nealy (2009a) in *Diverse Issues in Higher Education online* (3/19/09). In that article, the University of Maryland discusses goals for closing the racial graduation gap as Black students are graduating at lower rates than white and Hispanic students. Specifically, only 40% of Black students earn a degree within 6 years of entering college compared to 65% of all students and 70% of Hispanics. Even more troublesome is the fact that the disparity between Blacks and Hispanics has grown 10% (15 vs. 25%) over the past 3 years. Although Chancellor Kirwan states that Maryland is not unique in this area, and the article goes on to cite low retention rates in the state's historically Black institutions, I personally find it troublesome when high profile institutions such as the University of Maryland seem willing to compare themselves with others in an area such as this, i.e., Black graduation, while striving to be the absolute best in most other academic areas. In fact, a personal example of this was when I was testifying for a minority medical student who had filed suit for being dismissed from the University of Michigan. In response to one of my answers about the university's commitment to retention of minority students, the prosecuting attorney stated that those numbers were similar to other universities around the country. I replied "but does not the University of Michigan strive to be the best and is not usually satisfied to be "like the others"? So why are they willing to be just like other universities in this area?" As with most issues, it becomes a matter of priority and for many institutions, particularly majority universities, for many years, inclusion of minorities has not been a high priority. Another example of that is the University of Wisconsin which developed a diversity plan to increase the racial and socioeconomic diversity at its 26 college and university campuses, i.e., Plan 2008 (*Journal of Blacks in Higher Education* 2009a). The university a decade later admits that it "came up short of its goals" in that minority student enrollment went from 8.1 to 10.1% during those 10 years and also the graduation gap between Black and white students actually widened during that time to a current level of 23% points. One has to ask if other plans to improve the University failed as poorly, i.e., prioritization. They announced that a new diversity plan would be unveiled in Spring 2009. As still another example, the University of New Hampshire, after almost a decade spent recruiting and building a support base for minority students, remains a very White school in a very white state (Associated Press 2009a, "University of New Hampshire makes progress in increasing diversity"). The efforts to increase diversity were based in large part by a sit-in by students in 1998 demanding more diversity. Part of the objectives agreed to then was that there would be 300 Black students by 2004 and 10 Black tenure-track faculty by 2003. As of 2008, there were 197 Black undergraduates and eight Black tenure-track faculty members. Thus, as was seen with the University of Wisconsin, the "diversity" Plan did not work. Similar to Wisconsin, one of the reasons, in addition to the possible lack of prioritization of this objective within the institution, could be the environment as New Hampshire as a state is 95.5% white.

Causes for the Pipeline Leaks

The underrepresentation of minorities at all stages of the pipeline clearly represents a major concern, especially when one recognizes that with the changing demographics, a significant percentage of the K-12 population is represented by minorities. And, as mentioned, these leaks are not confined to one ethnic group. For example, in Texas where Hispanics are the fastest growing ethnic group, making up about 36% of the population, Hispanic students are falling behind educationally at an alarming rate. In fact, high school graduation rates and college enrollments for Hispanics lag behind those for both whites and Blacks (Associated Press 2009b, "Hispanic students lag in college admission"). In that article, it stated that only 68% of Hispanics in Texas graduate from high school in 4 years, which is 10 points below the overall rate, and just 42.5% of those who graduated in 2007 enrolled in college or technical training. As such, the leak in the pipeline between K-12 and college represents perhaps the most troublesome one of all as the "pool" is most assuredly there. Fixing this leak requires substantial changes in the K-12 system, relating all the way back to the *Brown versus Board of Education* issue of "separate but equal." And, perhaps now, the current economic crisis is the time to "fix the system." Jack Schneider (2009) from Stanford University wrote in *USA Today* (4/9/09) that perhaps now is the time to tackle education funding once and for all. He stated that just like at the time of the great Depression, school funding is dependent heavily on local property taxes, an inequitable funding formula that disadvantages the neediest children. As such, he pitches equal school funding, i.e., use federal aid to schools not only as a bailout but also as a means for encouraging them to create more equitable funding formulae. This would also require improved management and budgeting based on results but it could accomplish what President Obama wanted when he stated "The source of America's prosperity has never been merely how ably we accumulate wealth but how well we educate our people." Addressing this leak in the pipeline does not to take anything away from the importance of the leak in the pipeline at the faculty level but those numbers at the faculty and professional ranks will be discussed more in Chap. 8. It is important to point out at this point that regardless of why the numbers are so low, it is the fact that the numbers are low that presents many of the difficulties experienced by minorities at all stages of the pipeline. i.e., being in an isolated situation is not favorable for survival, let alone success and that is what ethnic minorities have experienced in academia and the sciences for many years. As a result, the numbers are extremely slow to increase and, with that, less people are available to provide support and mentoring for the others. At the undergraduate level, this has been the reason that Historically Black Colleges and Universities (HBCUs) were established and not only have remained but also have been most successful over the years (Burrelli and Rapoport 2008, "Role of HBCUs as Baccalaureate-Origin Institutions of Black S & E Doctorate Recipients"; Hannah 2009a, "Are HBCUs Still Relevant"; Nealy 2009b, "Education Secretary Duncan: HBCUs as relevant today as ever"), despite the fact that many have suffered from inadequate fiscal resources, especially when compared with Predominately White Institutions (PWIs). This problem of resources at HBCUs was addressed in an article by Errin Haines from the Associated Press, particularly in light of the economic woes of the country in 2009 (*Diverse Issues in Higher Education* 2/16/09). In the article, Haines focuses mainly on HBCUs which have traditionally been somewhat more financially sound than others, e.g., Morehouse, Spelman, Howard, and Hampton. Even in these cases, decreases in the amount of tuition due to losses in enrollment, coupled with less donations, have caused, in some cases, the phasing out of programs or schools as well as layoffs of staff as in the case of faculty of Clark Atlanta. Despite this, individuals quoted in the article such as Drs. Michael Lomax, UNCF President and John Franklin, Morehouse President are optimistic that the HBCUs will survive and flourish, especially based on their histories, and also on the fact that the HBCUs have always been segregated but not segregating (Rivers 2009, "HBCUs: Segregated but not segregating"). This is important to note since segregating institutions were, as defined by Dr. King, an instrument used by one race to uphold its political, economic, and/or social dominance through methods that work to the detriment of other races whereas a segregated institution is a body that is adversely affected by those policies. Another major impact of the economic downturn was seen with the announcement by the UNCF at the top of their Web site "The recession is hitting UNCF colleges hard. Hundreds may not graduate this spring," as reported by Stuart (2009) in *Diverse Issues in Higher Education* online (4/9/09). The UNCF, the usually somewhat low-key supporter of 39 of the nation's private, HBCUs is trying to raise as much as $5 million quickly so students ready to graduate in Spring can do exactly that. Dr. Lomax stated that he was confident that the organization could do it using emergency campaigns as well as corporate donors who have been long-time UNCF supporters. Still this represents another economic situation for education and, in particular, Minority Serving Institutions (MSIs). On the positive side of possible financial support of MSIs, the Lumina Foundation announced that it is requesting proposals for its Minority-Serving Institutions – Models of Success grant program (Staff 2009, "Minority-serving institutions called to serve as leaders in student retention"). The Foundation, recognizing that about one-third of all American students of color are educated at MSIs, which translates into almost 2.3 million students, states that these institutions "can serve as leaders in a national effort to improve college attainment rates." In direct comparison and contrast to funding at MSIs, many of the PWIs are research institutions and, as such, have access to many more financial resources.

Regardless, students from these institutions as well as other Minority Serving Institutions (MSIs) such as Hispanic Serving Institutions (HSIs) and Tribal Colleges are faced after graduation, in most cases, with once again, being much more isolated at an institution where they indeed represent the minority population. This is a major contributing factor among others why attrition is higher in these postbaccalaureate programs for women and minority students, despite a number of reports on the successful retention of minority students in PhD programs, e.g. Anderson 2008a, University of Texas program demystifies graduate school for minority students; Evans 2008, Mentoring magic: How to be an effective mentor-tips from two highly successful principal investigators; Glenn 2008, What universities can do to graduate more minority PhDs; Lederman 2008, Who produces Black PhDs? For those who do successfully make through these programs they are still faced with an isolation phenomenon in that the number of minority faculty members and health professionals, as mentioned above, come nowhere near those numbers that they experienced at the MSI or even their numbers in society. As such, they tend to congregate for support and to avoid the isolation, only to be labeled as segregationists. This is discussed in a most cogent way in the book *Why are all the Black kids sitting together in the cafeteria?* by Beverly Daniel Tatum (1997). Basically she states that this phenomenon is really not any different than the White kids or Asian kids who gather with others like them. Also, director John Singleton makes this point very well several times in the movie *Higher Learning*, which has its setting in an institution of higher education. The point is why does this seem to be a problem only when Blacks choose to do this? Still, despite the obvious reasons that this occurs and also despite the fact that it is most positive for those students involved as it strengthens their self-confidence and self-esteem, a recent article reported on a study that did show that "affinity groups" such as these can create more racial tension on campuses (Nealy 2009c, "Do affinity groups create more racial tension on campus?"). In fact, in the book *The Diversity Challenge: Social identity and Intergroup Relations on the College Campus* by Sidanius et al. (2008), the authors conclude that "intergroup contact reduces ethnic tension and increases in friendship across ethnic lines." Of course, at the same time, one still has to consider the very positive aspects of these affinity groups, particularly for those individuals who are part of these groups because they are in the minority within the environment. One possible solution to addressing this is to maintain the affinity groups but at the same time have the institution work to increase the interactions among the groups, thus allowing their independence but at the same time some interdependence upon one another. This is in contrast to an Arizona Legislative Committee which through an amendment of a state homeland security bill tried mandating the end of any such groups on campuses in their state. The result of an action such as this would be even more isolationism for the groups most affected leading to even more uncomfortable and unwelcoming environments (Metzler 2008, "Perspectives: Banning affinity groups shows lack of understanding") and using the banner of "diversity actions" for justification. Like all issues dealing with ethnic differences, serious efforts must be made to recognize the strengths of the differences and also to resolve problems attributable to the differences. This is particularly the case for those groups that are not well represented, such as certain ethnic minorities, and not just Blacks but for other underrepresented groups such as Hispanics and Native Americans as well. For example, in an article by M.J. Nealy, the life of a Native American on a PWI is chronicled (Nealy 2009d, "Chronicling the lives of Native Americans on Predominately White campuses"). From the article, it is seen that many of the same problems and concerns exist on these campuses for other ethnic minorities who too have been long underrepresented. In fact, it is not just about racism and ethnic minorities but also, as discussed in a *Diverse Issues in Higher Education* online article by A.R. Ford, about "ableism" or persons with disabilities (Ford 2009, "it's not just about racism, but ableism"). Mr. Ford, a Black man with a disability (muscular dystrophy), is a PhD candidate at the University of Pennsylvania. In the article, he discusses the issues that he has dealt with because he is a Black man in society and in academia but also states that there are times when his disability creates an "otherness" that overrides that due to his Blackness. As such, he believes that ableism should be viewed as a form of racism, even though he acknowledges that in many contexts intolerance that occurs against persons with disabilities is more permissible or palatable than racism based on color. Personally I can attest to that based on a statement made by an established scientist who was part of a committee for a national scientific organization. When the issue of the need for handicap facilities was brought up in a joint discussion with the Diversity committee, which was addressing disabilities as one of their emphases, the person said that "individuals who need facilities to accommodate their disability should not be in the laboratory or science. Why modify laboratory facilities to accommodate them?" Please note that this person was part of a committee for a national scientific organization that had close to 100,000 members world-wide, only serving to emphasize the concerns expressed by Mr. Ford. Also, it is interestingly but unfortunately not surprising, that the underrepresentation occurs in other disciplines besides science and although it is not the emphasis of this book to discuss these in detail, an article really brought this fact forward (Porro 2009, "Reviving the great debaters tradition"). This article discusses the fact that despite the success that the HBCU Wiley College had in the 1930s, which was actually chronicled in a movie "The Great Debaters," there were no HBCUs represented at the 2009 National Debate Tournament Championships (NDT), which is chartered and sanctioned by the American Forensics Association. Furthermore, the vast majority of the teams which were participating did not have a single Black debater. Again, although this is not a "science discipline," the lack of Blacks and HBCUs participating in the NDT demonstrates that an important source of leadership training is missing, one that directly ties to future Black executives for corporations, nonprofits, and government. It is all of these and, in particular, the latter, i.e., government, that plays a critical role in science by way of policy making, funding, and regulations. Still another example of an underrepresentation that is not in the sciences but is due to the same reasons as the low numbers in the sciences is that of law. Like in science, the recruitment of students of color to the legal profession has constantly been a source for discussion, usually in the context of "lowering the bar" to achieve greater diversity. As reported in *Diverse Issues in Higher Education online* (Warfield 2009,

"Perspectives: meeting the standards, not lowering them"), a recent law school graduate of color points to the availability of preparatory tools as the reasons that the numbers lag behind. Specifically, as cited also for those students of color, the author indicates that concentrating on mentorship, standardized test preparation, and critical reading/analysis skills during high school and college will vastly improve the law profession at all levels, i.e., admittance, retention, and employment. In fact, the author cites specifically the lack of preparation of students of color to "navigate the Law School Admission Council system, how to write effective personal statements, and how to prepare for the LSAT exam," which parallel almost exactly the problems that minority students in the sciences face at the graduate and professional school level. As he states, it is not a case of "pulling themselves by their own boot straps" as there are no straps. He goes on to recommend that similar efforts must be made to offer the same type of mentorship and guidance that is available for athletes, e.g., summer camps, sponsorship, mentoring, and tutorial programs. Finally, it is stated that "mentorship by academic and legal professionals will serve students of color much more than lowering the standard to be achieved". An example of such a program for the law profession is that of the City of New York School of Law's Pipeline to Justice Program (Hunt 2009, "Changing the face of the legal profession"). The aims of this program are exactly what is needed to address the issues raised in the previous article and, for that matter, to address similar issues in the other professions, e.g., medicine, so as to truly effect a change in diversity of the profession. This program has two main goals (1) to prepare law candidates to retake the LSAT and succeed in law school and (2) to allow students from diverse and underrepresented backgrounds a chance to enter the legal profession and serve their communities. These goals are accomplished through intensive practice test taking by the students and dedicated mentoring by the faculty. In addition, the administration of the program has taken on this responsibility most seriously, as evidenced by the comments of Dean Michelle Anderson. She states that "The Pipeline to Justice is making a terrific impact in terms of enhancing access to the profession. As gatekeepers to an overwhelmingly white profession, law schools have a responsibility to help the profession become more representative. [They] have to graduate students from [underserved] communities to move our country forward." Associate Dean Mary Lu Bilek agrees, stating "Society does not need more lawyers; it needs more lawyers who believe in justice, who understand the needs and concerns of communities that have been neglected or worse, and who understand the power and promise of those communities to move our country forward." Thus, whether it be science or law, the role of mentorship and individualization of efforts is crucial for the preparation of minorities for the next stage of their education and for their ultimate professions.

At the same time, despite the strengths and advantages of HBCUs and other MSIs, a recent data analysis from 83 federally designated 4-year HBCUs showed that only 37% of their Black students finish a degree in 6 years, which is actually 4% lower that the national college graduation rate for Black students. Most relevant to the issue of the shortage of Black men is the fact that just 29% of HBCU males complete their degree in 6 years. Granted, some schools such as Howard and Spelman have much higher graduation rates, certainly better than the national average, but many do not. For example, at 38 HBCUs less than 25% of the Black men who began college in 2001 had graduated by 2007! (Pope 2009a, "Under a third of men at Black colleges earn degree in 6 years"; Pope 2009b, "Men struggling to finish at Black colleges"). For some HBCUs, the figures were less than 10%. There are a plethora of possible reasons for these numbers. Walter Kimbrough, President of Philander Smith College in Little Rock, AR, puts the blame squarely on the HBCUs themselves, stating that they have become lazy and no longer offer the same type of nurturing, caring, "the President knows you" that was their trademark 40–50 years ago. As a result of his perspective on this problem, President Kimbrough was instrumental in establishing mentoring programs for men on the Philander Smith campus, programs which actually also actively recruited the students to the programs. Additionally, it is President Kimbrough's belief that since many of the men come to campus with the idea that they should not ask for help, the answer is for the institution be more intrusive. Graduation rates have indeed gone from the teens to near 30% as a result of these collective efforts. Other reasons, in addition to the reluctance to ask for help, and which have been mentioned already, some of which are not unique to HBCUs, include (1) the lack of role models, (2) lack of preparation, and (3) finances. The last one is of special importance since the lack of sufficient resources and facilities is all too characteristic of many HBCUs and, as such, contributes significantly to this problem. As an example, Philander Smith is not a campus with significant resources but at the same time is relatively small in terms of student numbers and as a result can give more attention to the individual student. Still, as stated throughout, money and financial support is critical and a school like Philander Smith is hit hard by the problems as cited above that the UNCF is experiencing. At the same time, a school like Philander Smith is certainly "eligible" to submit a proposal to the Lumina Foundation for funding as a Model of Success institution (Diverse Staff 2009, "Minority-serving institutions to serve as leaders in student retention"). The difficulty that a school such as Philander Smith has, as well as most other MSIs, in submitting grants is the lack of research infrastructure and faculty time. Most faculty members at MSIs teach an equivalent of four courses per semester and when that time/effort commitment is combined with the lack of research infrastructure as a teaching institution, it is difficult to prepare grant submissions even when the agency has specifically identified their institution as the type for funding. Once again this goes back to monies as both "release time" and the development of a research infrastructure depend upon resources. As a personal note, I presented a career development seminar at Philander Smith in 2009 and did find a very active engagement of Black males at that time.

Shortage of Mentors

Directly related to the scarcity of the numbers is of course the lack of mentors which obviously has a very detrimental effect on the retention of the individuals at all stages of the pipeline (Bass et al. 2008, The University as a mentor: Lessons learned from UMBC Inclusiveness Initiatives; Evans 2008, Mentoring magic: How to be an effective mentor, tips from two highly successful principal investigators; Fain 2007, Still young enough to be hungry; Glenn 2008, What universities can do to graduate more minority PhDs; Lederman 2008 Who produces Black PhDs?; Monastersky 2006, Brief intervention improves achievement of students subject to negative stereotyping, study finds; Nealy 2008a, New report highlights schools that make minority student success a priority; Yue 2008, Bonds to faculty help keep Latinos in STEM majors. To that point, in a recent article in *Diverse Issues in Higher Education*, Dr. Ansley Abraham (2009) sees the increased enrollment of Blacks and Hispanics as an opportunity for junior faculty to mentor minority students to graduation and into the faculty ranks. To do this, however, he promotes the training of young minority faculty members in the area of mentoring through groups such as the Southern Regional Education Board (SREB) – State Doctoral Scholars Program as well as the Compact for Faculty Diversity's Institute on Teaching and Mentoring (2008). At the same time, he recognizes and states that the depth of the problem in that if every one of the Black, Hispanic, and Native American students who received a PhD in 2006 chose an academic career, there would still not be enough candidates to ensure that every college or university in the country could hire just one minority professor! Unfortunately, very often possible solutions or at least efforts in an area such as this are only made after a situation has surfaced as in the case of Massachusetts Institute of Technology (MIT) and the case of James. L. Sherley. In 2007 Dr. Sherley began a hunger strike outside the Provost's Office alleging racism in his tenure denial. At the time only 27 of MIT's 740 tenured faculty were Black, Hispanic, and Native American (that is 3.6%). Dr. Sherley, despite various efforts after the strike, including an EEOC investigation, the breaking of ties with MIT by Frank Douglas, a Black professor and Bernard Loyd, a Black former trustee and a letter authored by 11 MIT faculty members outlining a number of problems with the case, never did receive tenure (Hennick 2009, "Creating a sustainable pipeline"). Although Professor Douglas was not a signer of the letter, from his interactions with other faculty, he was quoted as saying "what I discovered…is that many of the young (minority) faculty were unsure as to how they would be evaluated and what type of career they would have." This is similar to the results of recent articles that discuss the issues that young minority faculty raised regarding tenure and their positions (Jaschik 2008, "Racial Gaps in Faculty Job Satisfaction," *Inside Higher Education*; June 2008 "On the road to tenure, Minority Professors report frustrations," *The Chronicle of Higher Education* 12/5/08). As a result of this case and the surrounding events, the institution (MIT) indicated a commitment to ensure a welcoming environment for faculty of color and as of 2009, 34 faculty of a total of 767 are underrepresented minorities (4.4%). This commitment included the establishment of new positions that focus on faculty and staff diversity, including two associate provosts for faculty equity, as well as other initiatives including an event called the Diversity Leadership Congress which brought together administrative, faculty, and student leaders. In response to this and the other efforts, including a new Diversity Web site (Journal of Blacks in Higher Education 2009a, b, "MIT debuts new diversity Web site") directed toward more minority faculty at MIT, the overall objective, as stated by the Provost, is to focus on creating a "sustainable pipeline" rather than to be as concerned with the current numbers. There is no question that a sustainable pipeline is critical in making a difference and time will only tell whether this indeed happens at MIT. Regardless, despite its outstanding ranking among top institutions in the country in the area of science and technology, MIT is not able to claim leadership on issues of diversity. Unfortunately, this is in no way unique to MIT but rather, in general, there continues to be a continuing racial shortfall in tenure rates. Other examples. i.e., Duke, Harvard, Michigan, and Virginia Tech were included in an article entitled "Whatever happened to all those plans to hire more minority professors" which appeared in *The Chronicle of Higher Education*, written by B. Gose (2008). As with MIT, the successes of these outstanding institutions with regard to minority faculty recruitment ranked nowhere near their top rankings with regard to such things as research, funding, etc. In fact, some of the successes that Virginia Tech had seen earlier, due in large part to the commitment of an Associate Dean, who had since left, were affected negatively due to a "reining in or elimination" of some of the tactics that had worked. More recently, even the existing diversity language at Virginia Tech, relative to tenure and promotion, has now been challenged by the Foundation for Individual Rights in Education (Wilson 2009, "Critics challenge diversity language in Virginia Tech's tenure policy"). As a result, the ground that had been gained by at least one university that had instituted some successful tactics was lost and may not be regained. And, as we discuss later in Chap. 8, concerns about tenure with minority faculty are paramount, especially considering the paucity of underrepresented faculty in general and at the tenured stage. For example, *The Journal of Blacks in Higher Education* (2009d) reported data from the Department of Education showing that in 2007 Blacks made up 3.4% of all full-time faculty members at degree-granting institutions in USA. And, of all tenured faculty members, the number of Blacks was 4.6%; however, that number of tenured Black faculty only represented 35% of the full-time Black faculty as compared to 44.6% of all white faculty who had tenure. Dr. Metzler provides some interesting perspectives on the tenure process as it related to minority faculty (*Diverse: The Academy speaks*, Metzler 2009, "The case against cultural standardization in tenure decisions") which will be discussed in Chap. 8.

Impact of Antiaffirmative Action

Moreover, this is not a new phenomenon; in fact, efforts have been directed toward addressing this issue for many years, e.g., two of the major federal funding sources, i.e., the National Institutes of Health (NIH) and the NSF have had a number of designated programs for addressing underrepresentation for many years. In addition, a number of private foundations have also provided support for many years to address this disparity as well, e.g., Sloan, Gates, Ford, etc. Although these programs can be deemed successful, they have only scratched the surface with regard to the total numbers, which is particularly significant with the current changing demographics that USA is facing. The persistence of this problem is of course due to a number of factors, including the lack of educational access and preparation, institutional racism, cultural issues, and attitudes against "affirmative action," especially in recent years. These efforts by a number of groups, e.g., Center for Individual Rights and Center for Equal Opportunity have indeed decreased the numbers, and effectiveness, of minority-targeted programs, that as mentioned, had been successful over the years. Major examples of the results of actions by these groups include Prop 209 (CA), Prop 2 (MI), and similar actions in Texas and Washington. However, some changes have begun to be seen even with this movement, as evidenced by the decision by the California appellate court to uphold a school assignment diversity plan that was instituted in Berkley, CA (Matthews 2009, "Perspectives: a win in the diversity column"). In this case, it was decided that the plan did not violate Prop 209 since it considered the racial makeup of the neighborhoods, not the characteristics of the individual students. As a note at this point, it is interesting that other examples of targeted admissions programs, such as legacy, i.e., a relative previously attended the institution and athletic recruitment have not been attacked, or even challenged, as to their "fairness" such as has been the case with those programs that have targeted individuals that have been traditionally underserved and almost excluded in areas of education such as science. Perhaps that has been due to the fact that the groups that have supported preferences such as legacy, but attacked the minority programs, have quite often been members of those former groups and therefore had the "privileges." An excellent source for such discussions is the book, Color and Money by P. Schmidt (2007). Interestingly, it has now reached a point where at least one school has decided to have an "orientation for whites" (Jaschik 2009a, "Orientation for whites"). Mount Holyoke College plans this year to start a special session for white students as part of a new orientation program titled "Promoting Intercultural Dialogue and Creating Inclusion." There are mixed feelings about this as some believe that it will lead to more dialog between minority and white groups whereas others believe that it will lead to more segregation. There is no question that more dialog is necessary in the area of minority issues, especially relative to white privilege. Time will tell if this orientation for whites experiment works to accomplish that as currently it seems to be the only program like it in the country. Regardless of what programs exist, the one parameter that most often determines success vs. failure in almost all efforts is money, whether it is at the governmental level, at an institutional level, or at a personal level. Unquestionably, a major part of the difficulty in addressing this problem is that there has just not been enough funding directed toward these programs to adequately address the gravity of the problem, when considering the total numbers. Unfortunately until that changes, the type of changes that are needed will continue to be delayed, if not totally thwarted. As a positive step in that direction, the earmarks as part of the economic stimulus package under President Obama include HBCU historic preservation monies as well as specific designations for HBCUs, HSIs, and Tribal colleges. Moreover, these funds are in addition to monies that are provided for Department of Education programs such as TRIO, GEAR UP, and K-12 Title 1 (Dervarics 2009, "Earmarks helpful for minority-serving institutions struggling in tough economy").

Financial and Racism Considerations

Interestingly, the link between access to education and financial considerations is quite direct. For example, we know that despite the Brown v. the Board of Education decision, education in our country still represents a most segregated institution. Why? Public schools are supported by taxes and, as such, schools in well-to-do communities are going to have better facilities and often better teachers than schools in more financially deprived neighborhoods, which are very often either urban or truly "rural." And, more often than not, ethnic minorities often attend schools with less funding and resources; hence the basis for disadvantaged and underserved individuals, often both financially and educationally. Additionally, the relationship between "social class and college readiness" is significant and discussed in an article by Patrick Sullivan (2009) in *Academe Online* 1/20/09 entitled "An open letter to ninth graders." As that issue relates to this text, unfortunately in addition to not usually being as "college ready" due to both their K-12 experiences and lack of exposure in the home, these students often also do not have access to mentors as those from the well-to-do schools, i.e., often lacking both natural and planned mentoring. This issue is one that our society must address by making education a higher priority; one that will allow for equity and equality regardless of the financial status. However, in the meantime, it is essential for those students to have access to mentoring by any means necessary. Individuals as well as groups (such as those mentioned in Chap. 2) help to serve this purpose. This carries over directly to the advice that the NSF got from Hispanic Serving Institutions (HSIs) through the Hispanic Association of Colleges and Universities (HACU) in response to the question of how to tackle the dearth of Latinos in the science, technology, engineering, and mathematics (STEM) fields. The message was clear, as stated in the article in *Diverse Issues in Higher Education* 3/2/09 by Branch-Brioso (2009), i.e., "pay Hispanic students to do research or you will never get them, and keep them, in STEM fields, as they are working class students and need the financial support."

The inequality that is displayed in education represents an aspect of institutional racism that too has to be addressed in our society, whether the institution is a university, a corporation, a government agency, or the local supermarket. There are certain state and federal guidelines to prevent discriminatory behavioral actions; however, there has to be accountability for these guidelines and policies to be enforced and in recent years the number of these cases handled by the Office of Civil Rights (OCR) had fallen precipitously, despite a significant increase in job discrimination claims filed with the EEOC in 2008 (Hannah 2009b, "Job discrimination hits record high: is economy to blame?"). Perhaps with the struggling economy, there is a movement to acknowledge more biases associated with the job, only time with tell if these translate into OCR claims. Regardless, this issue of racism and discrimination can still continue to be addressed individually through mentors who choose to identify and mentor those students who traditionally do not have ready access to mentors. In doing so, these individuals, especially those who are not minorities, help to address the racism, not only at the level of the individual student, but also, institutionally, by recognizing the importance of different cultures. Ultimately it will take these efforts in combination with those of offices such as OCR and EEOC to truly make the difference needed.

Role of Community Colleges

It is significant at this point to discuss the role that community colleges are playing in the educational pipeline, especially for minority students. Historically, community colleges had a poor reputation, i.e., students went there because they could not get into 4-year institutions or they only wanted to learn a trade or they wanted to play athletics. This is not to say that some enrollees did not go to community college for those reasons; however, there were many other very good reasons that students chose community college, such as lower costs, academic strengthening, and geographic convenience. For these, as well as other reasons, community colleges as a group represent the most ethnically diverse educational institutions in the country. For example, 47% of Black and Asian students, 55% of the Hispanic students, and 57% of the Native American students are enrolled in the almost 1,200 community colleges across the country (Waiwaiole and Noonan-Terry 2008, "Perspectives: Need to equip, prepare community college faculty has never been greater"). Moreover, the diversity at community colleges extends into other areas, e.g., 47% receive financial aid, 59% are women, 29% are the first in their family to attend college, and 50% of the part-time students work full time. Also very importantly, 42% are enrolled in at least one remedial reading class. An additional factor that is to be considered with regard to community colleges and the educational future of USA is the anticipated boom in "baby-boomer students" (Hoover 2009, "Community colleges anticipate boom in baby-boomer students"). At the American Association of Community Colleges 2009 annual convention, Mary Sue Vickers, the director of the *Plus 50 initiative*, told the convention attendees that community colleges must do more to engage older students and prepare them for jobs. In fact, this initiative is a 3-year project that was designed to create and support programs for adults over 50 at 15 community colleges, and despite the fact that a high percentage of community colleges surveyed had offerings specifically for older students, only a little over half of those had work-force training and career-development services tailored specifically for them. As a result of this survey, the Director stated that "society's ideas about aging have not kept pace with reality or with how baby boomers see themselves." Statistics support this statement as the Bureau of Labor Statistics projects from 2006 to 2016, the number of workers aged 55–64 will increase by 36.5% while the number of workers who are 65 and older will increase by 81%. Along with this from the community college perspective comes the marketing issues as putting a photo of an 18-year old on a community college brochure is not a good recruitment tool for someone who is significantly older than the person in the brochure, i.e., community colleges have to produce more relevant and effective recruitment materials. Another example of how community colleges are meeting the needs of society is the rise in distance education enrollments. At the 2009 annual meeting of the American Association of Community Colleges, it was reported that distance enrollments grew almost 30% in the past 2 years with the concern that these types of increases could not be sustained indefinitely, i.e., funding of these programs (Jaschik 2009b, "Rise in distance enrollment"). This concern may be exacerbated by the fact that funding in general is becoming more and more limited plus the announcement by the Sloan Foundation that it is closing its online-education grant program, a program that has funneled approximately $80 million since the early 1990s (Parry 2009, "Sloan foundation ends major grant program for online education"). Once again it is observed that funding represents the critical requirement for ensuring that the education process is successful, in this case at the community college level. This was in fact a topic of major discussion not only by a panel of "first" presidents at the 2009 American Association of Community Colleges meeting but also throughout the conference as community colleges have to "get on board" relative to fund raising just as 4-year institutions have done over the years. Finally, with respect to community colleges adapting to the needs of society, a number of unemployed are now seeking training for "green collar" jobs, such as installing solar panels, repairing wind turbines, and other work related to renewable energy (Chea 2009, "Unemployed seek training for 'green collar' jobs"). To respond to this demand, community colleges are using monies from the federal stimulus package. As examples, Palm Beach Community College in Florida is offering a new associate degree program focusing on alternative energy sources, while Central Carolina Community College in North Carolina has long wait lists for green building and renewable energy classes. In Michigan, which has been hit as hard as any states with regard to the economic downturn, Lansing Community College has seen an enrollment growth in its alternative energy degree program from 42 students in 2005 to 252 students in 2008. As a most positive note, with the current recognition of community colleges as an integral part of postsecondary education and the support throughout President Obama's administration, it is hoped that sufficient funding will be identified to fund the needs of the community colleges, especially as they make the necessary adjustments to meet society's needs.

Unquestionably, community colleges represent a critical step in the educational pipeline, especially for minority students. As a result, a number of efforts have been directed to facilitate, improve, and in some cases revamp community college education. Examples include President Obama laying out a community college plan in his campaign (Associated Press 2008a, "Obama lays out community college plan at campaign stop"), a community college working with the community unemployment issue (Miranda 2008, "How one community college gave displaced factory workers the confidence to enroll"), the relationship between community colleges and the economy (Ashburn 2008a, "Community colleges are key to shoring up the US economy, report says"; Associated Press 2008c, "Weak economy spurs growth for community colleges"; Associated Press 2008d, "Gates foundation to invest in community colleges"; Jaschik 2008d "Gates Foundation to spend big on community colleges") as well as other efforts to promote community college enrollment (Associated Press 2008e, "Wisconsin officials consider community college baccalaureate plan"; Jaschik 2009c, "Morphing community colleges"; Pluviose 2008, "More high-achieving students are choosing community colleges first"; Nealy 2008b, "New study looks at retention in community colleges"; Anderson (2008a), "Philadelphia program offers free tuition for community college students near degree completion"; Goodall 2009, "Bridging the gap from GED to community college student"). Furthermore, *Diverse Issues in Higher Education* developed a partnership with the National Institute for Staff and Organizational Development (NISOD), which is a consortium of more than 700 community colleges to provide a monthly column focused on community college issues. Also, with this emphasis on community college education, problems that had been seen such as student isolation/fear in a new surrounding plus miscellaneous administrative hassles have been addressed through various efforts (Ashburn 2008b, "Fears and Administrative hassles deter community-college students in their first term, they say"). Another example was a result of the Survey of Entering Student Engagement (Sense). In this case, a call to meet new students at the front door represented a straightforward solution (Sander 2008, "At community colleges, a call to meet new students at the front door"). In another case, the combined efforts of the Community College Survey of Student Engagement and the NISOD resulted in the attainment of student data that was then used to address issues regarding the students' community college experience (Arnsparger 2008, "Student input helps community colleges improve overall excellence"). In still another case, using the Posse Foundation success model of peer studying as a basis (Reynolds 2008, "Riding into college with your posse"), the nonprofit research organization MDRC tracked the progress of freshman at Kingsborough Community College of CUNY and found that taking courses as a group helped these community college freshman succeed (Supiano 2008, "Taking courses as a group helps community college freshman succeed, study finds"). Still other reports demonstrated that mentoring programs help keep students engaged (Hayes 2008, "Community college mentoring programs help students stay engaged") and that Black students are among the most engaged in community colleges (Nealy 2008c, "Report: Black students among the most engaged at community colleges"). Recently, Dickinson College, in Pennsylvania, developed a comprehensive transfer partnership with four local community colleges that goes much beyond the typical articulation agreement (Moltz 2009b, "New approach to community college transfers"). With this new initiative, if transfer students meet the academic criteria, e.g., 3.25 GPA, they are provided a merit scholarship, i.e., $10–15k, in addition to any need-based aid that they qualify for. For students early in their community college experience, an interest in Dickinson is "rewarded" by advising, counseling, and visits to the campus. Because financial support is a major component of this program, it is expected that it will increase not only the successful transfers but also the diversity of the transfers. The bottom line from all of the efforts cited is that attending a community college can be a most positive component in a student's educational pipeline, especially if there are good reasons for doing so.

And, as mentioned, for the under represented ethnic minority students there often are such reasons, including the economic situation in the country in 2009. This "recession" is responsible for sending even more students to community colleges (Tirrell Wysocki 2009, *Diverse Issues in Higher Education*, "Recession sending more students to community colleges"). In this article, the author cites the fact that the average annual cost of community colleges is $2,402, whereas it is $6,585 for tuition and fees for an in-state resident at a public 4-year institution and $25,142 for a private 4-year school. Also, in response to the economy, in some cases community colleges are offering free tuition to the unemployed (Delos 2009a, *Diverse Issues in Higher Education* 2/13/09). However, as a result of such occurrences, the influx of students is creating problems at some schools as they are just not able to accommodate a significant number of new enrollees, which may actually lead to community college turning students away (Delos 2009b, "Community colleges may soon start turning students away"; Selingo 2009, "Community-college leaders confront a challenge: enrollments are up but money is not"). On the positive side, due to the fact that academia has now, for the most part, recognized the importance of community colleges and the education that they provide, students can use that experience to their advantage in progressing up their educational ladder, something that was difficult for them to do previously. There is no question that there is a "community college surge" (Jaschik 2009d, "Community college surge") that has resulted in turn in an increase in almost every major type of program offered. It is hoped that these institutions can continue what they have done for so long, i.e., "doing more with less."

Another major step in acknowledging the role of community colleges nationally was the recommended appointment of Chancellor Martha J. Kanter from the Foothills De Anza Community College District as the US undersecretary of education, which is the second- or third-highest ranking position in the Department of Education (Lederman 2009, "2-year college leader is US nominee"; Field 2009a, "Obama pick shows focus on training work force"). In this position, Ms. Kanter oversees policies, programs, and activities related to postsecondary education, vocational and adult education, and federal student aid. Unquestionably, the appointment spoke volumes as to the importance placed on community college education by the presidential administration. In fact, in response to the appointment, Gerardo de los Santos, President and CEO of the League for Innovation in the Community College stated "But in the broader scope, it also says something about the meaningful role that community colleges are playing in our society to step up and provide economic development, revitalization and educational access." Also, George R. Boggs, President of the American Association of Community Colleges said "I think that this administration realizes that community colleges are a real unrecognized workhorse for education and work-force development." Finally, Secretary Duncan further confirmed the role of community colleges by stating "community colleges are a vital, vital part of our postsecondary education system and an extremely important part of restoring our economy and ensuring our students can compete." Following the Kanter appointment was another appointment recognizing a community college administrator. Glenn Cummings who currently was Dean of Institutional Advancement at Southern Maine Community College and formerly speaker of the Maine House of Representatives was appointed to become deputy assistant secretary of education, a post in which he will work to get more Americans to enroll in college (Field 2009b, "Maine community-college dean is named federal higher-education post").

Along with these positive actions regarding community college education, a somewhat more sobering statistic comes from a look at the ethnic make-up of community college trustees, especially considering the demographics of the community college student population and the fact that the trustees govern the community colleges. The Association of Community College Trustees (ACCT) unveiled results of a comprehensive survey among almost 750 local boards from 39 states and 34 state boards (Moltz 2009c "Who are community college trustees?"). Their data showed that 82% were white, 9% were Black, 4% were Latino, and 2% were Asian. Other statistics showed that this group was 66% male and 34% female, more than 50% made more than $100,000 annually, and only 29% were from the education profession. Again, considering the group of students that are served by community colleges, i.e., 50% or more are from each of the underrepresented ethnic groups, approximately 50% receive financial aid and almost 60% are women, having trustees that might relate more closely to that population could possibly better address some of the issues that community colleges are currently and will face in the future.

A final issue that has to be noted and addressed is the perception of community college courses and the students that matriculate in them. "Junior" colleges in the past were seen as places where students could go for furthering their athletic abilities or places where students went for a trade as opposed to where they went for legitimate reasons including preparing for a 4-year institution where they could eventually graduate. Although that perception is not as wide spread as it once was, especially in academia, it is still much more pervasive than it should be. And, as so often is the case, media unfortunately can potentiate this misperception. For example, a few years ago Jay Leno, on his Tonight show, made some comments about community colleges that were not only disrespectful and derogatory but also not based on facts. Although an advocate for community colleges was able to ultimately meet with Mr. Leno and clarify some of the misperceptions, many who heard his original comments never heard the follow up. With regards to the issue of remediation and its perception related to community colleges and the minority male, an article by Moltz (2009d) (*Inside Higher Ed* "Black men and remedial education") makes some very cogent points. Discussions from a meeting of the American College Personnel Association, which represents student affairs administrators, focused on remediation at both community colleges and 4-year institutions, emphasizing the paucity of the Black male in education in general and in science in particular. Several of the attendees defended strongly the role of remediation at 4-year institutions, basing their comments on a study which analyzed the effect of a particular developmental program at an HBCU on retention and persistence of Black males. Ivan Harrell from J. Sargeant Reynolds Community College in Virginia suggested that due to various impairments that Black men encounter in high school they are more likely candidates for remedial education and as such would be more affected by possible cut backs in support of those programs, both for pure financial constraints but also for perceived lack of effectiveness. He specifically cited at least 22 states that have either educed or eliminated remedial coursework from their public institutions, including some HBCUs. His thoughts were that some of this was due to the public perception, i.e., the institutions needed to increase their public image/prestige and eliminating remediation would help to do that by making the students go to community colleges for that. As for specific data, Robert Palmer, a professor at the State University of New York (SUNY) at Binghamton presented results from a study at an HBCU where a group of Black men were followed from the time that they entered a remedial program as freshman through graduation. Based on extensive interviews, he found that the young men "warmed" to the idea that the program indeed offered them a "second opportunity to earn a college degree" as well as providing them with self-confidence and self-esteem. Despite this study being more qualitative than quantitative, it strongly supported the concept that remedial education very often enhances "sound and academic integration," especially for minority students. So, although remedial education may not be maintained at all 4-year institutions, as mentioned elsewhere in this text, or, in fact, may be on the chopping block at some, undoubtedly by developing "varied measures of assessment," schools could justify these programs and their benefits. As a personal note from my experience regarding remediation at my 4-year institution, i.e., CSUDH, about 90% of our incoming students need math and English remediation, very much due to the school system in the area of South Los Angeles where the university is located. Despite this fact being well known, many faculty members complain that they should not have to teach remediation. The obvious response is that there is a need for it to be taught and as educators at the school it is our responsibility, certainly until such time that the local K-12 schools can provide the appropriate education so that is no longer needed, or at least needed only minimally at the college level. Short of that happening, relationships between secondary, i.e., local high schools, and postsecondary institutions, CSUDH, must be strengthened so as not to eliminate some important educational opportunities, especially for minority males who are already negatively impacted in the educational system. One additional approach that addresses both the introduction to STEM as well as remediation at the community college level is the efforts expended at Eastfield College, which is part of the Dallas County Community College District. The College reported at the League for Innovation in the Community College 2009 meeting (Moltz 2009e, "Beyond 'drill and kill'") that, with support from the NSF, "they had increased the number of STEM majors of URMs by 57% while at the same time raising the retention rate from 15 to 46%." They accomplished this by "rethinking the way that the students are introduced to science" and using stipends to pay the students to work in a science environment, rather than throwing them into an introductory physics or biology course to experience "drill and kill." These efforts have been complimented by bringing in featured scientists to talk about careers (they indicated that the free pizza did not hurt the attendance!). They are now moving in a direction to address the remediation, especially in math since most require it and, as such, are held back and even discouraged from moving forward into science disciplines. Math remediation will be offered by specific components, e.g., fractions, an example where the student may be weak. Also plans are to offer the remediation in the summer so that they can enter college "math ready." It is obvious that educational efforts at the K-12 level and the college level, i.e., both community college and 4-year institutions must make efforts to address weaknesses wherever they occur and also develop ways to remedy them.

With all of the past concerns related to community colleges, including misperceptions and remediation education being two of the major ones, it has only been in recent years that academicians, particularly at research intensive universities, and only at some, are beginning to recognize the role that community colleges are playing and will continue to play in the future. This is especially important not only as society adjusts to the educational needs especially in light of the country's economics but also since these misperceptions can play a significant role later in the student's pursuit of further education. For example, admissions decisions for professional schools, e.g., medical schools, need to know how to interpret information that is common to community colleges and the 4-year institutions where they have articulation agreements, but not common to many research intensive and/or professional schools, i.e., articulated courses accepted at 4-year institutions should be considered the same as courses taken at the 4-year institution. If Admissions personnel are not aware of these nuances associated with community colleges and more students, especially minorities, are coming forward with credentials from community colleges, addressing the underrepresentation will once again take a major hit. This problem can actually be exacerbated by the fact that oft times the "prehealth advisors" at community colleges are not as well prepared concerning professional school admissions practices as those at 4-year institutions plus often they may not even believe that the advice is really theirs to give since the student will be getting many of the prehealth requirements at the 4-year institution. Similarly, there will often be an even greater absence of mentors at the community colleges who are knowledgeable about these admissions practices at professional schools. Thus, since the underrepresented numbers in the graduate and health professions are already an embarrassment, and more minorities are attending community colleges, admissions personnel at the graduate and professional schools have to become educated as to interpreting community college transcripts. Getting past the stereotyping and the misperceptions can be the first productive step.

Certainly the emphasis on community colleges as demonstrated by the Obama administration can help tremendously in this regard, not only as mentioned earlier the President's statements about community college education and his appointment of community college leaders to key education posts, but also the fact that Dr. Jill Biden, the wife of the Vice President, is currently teaching at a community college and even more to the point, her thesis was on "how to retain students in community college." To this point, she accompanied the US Secretary of Education Duncan on a visit to Miami Dade College, the largest and most diverse community college in

America as well as the one that awards more associate degrees to Hispanic and Black students than any other college in the country. During the visit, she and the Education Secretary spoke individually with students as well as college administrators (Farrell 2009, "Duncan: Community colleges important to restoring economy"). As part of those discussions, it was pointed out that part of the stimulus package would aid education through increases in Pell grants and Work-study awards as well as through tuition tax credits for families that would benefit most relative to community college attendees. These types of increases lend support to Secretary Duncan's statement that "improving education is the civil rights issue of our generation." Community college education is an excellent place to start and the support for the community college education throughout the President's administration bodes well for that happening.

Thus, in the meantime, and until the size of the pool can be significantly increased at all stages of the pipeline, there are efforts that can be made to "work with the pool that is available," and the most important of these is the individualized and personal mentoring of the students. It has been reported that this type of effort is most effective to the success of minority students. Thus, even without the benefit of a funded program specifically designed to accomplish this goal, efforts by individuals can result in a similar effect. Not surprisingly, this is most effective through personal one-on-one interactions; however, as already mentioned, the use of communications through email and Web chats can also serve as a very meaningful exchange. In fact, when this is coupled with providing guidelines and directions to the students, the need for the direct personal interaction is ameliorated (but never eliminated!). To do this most effectively in the academy, institutions must truly recognize the importance of mentoring and as such provide support for it, whether it is as additional salary, release time, or anything else that provides the faculty member with the type of support necessary to say "mentoring is important to our institution."

Increasing the numbers of minorities into the sciences is definitely a "win–win" situation, especially considering the fact that by 2043, whites will comprise less than 50% of the population, with Latinos representing 30% (Visconti 2009, "Can companies survive without a concern for diversity"). Currently the "majority is the minority" in California and Texas. As a result, if the workforce is to mirror society, then addressing the underrepresentation of minorities now, especially in the areas of science, technology, and health, is a must. The positive aspect of this issue is that we know where the problems exist, i.e., leaks in the pipeline, and actually how to fix them. Unfortunately these "fixes" require funds and that is why the problem of underrepresentation has not only persisted, as mentioned, but has gotten worse. Short of the allocation of monies to fix these leaks, both now and in the future, efforts such as writing this book are being undertaken to provide assistance in areas where we know the problems exist. Additional references that provide insights into this area include Adams 2002, Get up with something on your mind; Jones 2006, Just what the PhD ordered; Moore & Penn 2005, Finding your North; National Academy of Sciences 1997, Adviser, teacher, Role Model, Friend on being a mentor to students in science and engineering; National Science Foundation 1997, Mentoring for the 21st century workforce.

References

Abraham A (2009) Perspectives: A golden opportunity. Diverse Issues in Higher Education online http://www.diverseeducation.com/artman/publish2009/01/20

Adams HG (2002) Get Up with Something on Your Mind, Gemstones for Living, Marietta, GA

Anderson MD (2008a) University of Texas Program Demystifies Graduate school for minority students. Diverse Issues in Higher Education online http://www.diverseeducation.com/artman/publish/2008/09/04

Anderson MD (2008b) Philadelphia program offers free tuition for community college students near degree completion. Diverse Issues in Higher Education online http://www.diverseeducation.com/artman/publish/2008/08/13

Arnsparger A (2008) Student input helps community colleges improve overall excellence. Diverse Issues in Higher Education online http://www.diverseeducation.com/artman/publish/2008/04/03

Ashburn E (2008a) Community colleges are key to shoring up the U.S. economy, report says. The Chronicle of Higher Education http://chronicle.com/daily/2008/01/1464n Accessed 1/31/08

Ashburn E (2008b) Fears and administrative hassles deter community college students in their first term, they say. The Chronicle of Higher Education http://chronicle.com/daily/2008/03/1919n Accessed 3/4/08

Associated Press (2008a) Obama lays out community college plan at campaign stop. Diverse Issues in Higher Education online http://www.diverseeducation.com/artman/publish/2008/02/21

Associated Press (2008b) Wisconsin officials consider community college baccalaureate plan. Diverse Issues in Higher Education online http://www.diverseeducation.com/artman/publish/2008/02/28

Associated Press (2008c) Weak economy spurs growth of community colleges. Diverse Issues in Higher Education online http://www.diverseeducation.com/artman/publish/2008/08/22

Associated Press (2008d) Gates Foundation to invest in community colleges. Diverse Issues in Higher Education online http://www.diverseeducation/artman/publish/2008/12/10

Associated Press (2009a) University of New Hampshire makes progress in increasing diversity. Diverse issues in Higher Education http://www.diverseeducation.com/artman/publish/2009/04/09

Associated Press (2009b) Hispanic students lag in college admissions. Diverse Issues in Higher Education online http://www.diverseeducation.com/artman/publish/2009/04/07

Bass SA, Rutledge JC, Douglass EB, Carter WY (2008) The University as mentor: Lessons learned from UMBC Inclusiveness Initiatives. Council of Graduate Schools: Volume One in the CGS occasional Paper Series on Inclusiveness

Branch-Brioso K (2009) What will it take to increase Hispanics in STEM? Money, of course. Diverse Issues in Higher Education online http://www.diverseeducation/artman/publish/2009/03/02

Burrelli J, Rapoport A (2008) Role of HBCUs as Baccalaureate-Origin Institutions of Black S & E Doctorate Recipients. *InfoBrief SRS* August 2008

Chea T (2009) Unemployed seek training for "green collar" jobs. Diverse Issues in Higher Education online http://www.diverseeducation.com/artman/publish/2009/04/14

Chubin DE (2008) Strategies for effective diversity programs in graduate schools 2008 Council of Graduate Schools (CGS) Meeting 12/03/08

Compact for Faculty Diversity's Institute on Teaching and Mentoring (2008) Mentor, Friend – or Both?

Delos RC (2009a) Recession: Some community colleges offer free tuition to unemployed. Diverse Issues in Higher Education online http://www.diverseeducation.com/artman/publish/2009/02/13

Delos RC (2009b) Community colleges may soon start turning students away. Diverse Issues in Higher Education online http://www.diverseeducation.com./artman/publish/2009/02/19

Dervarics C (2009) Earmarks helpful for minority-serving institutions struggling in tough economy. Diverse Issues in Higher Education online http://www.diverseeducation.com/artman/publish/2009/03/13

Evans J (2008) Mentoring magic: How to be an effective mentor; tips from two highly successful principal investigators. The Scientist 22(12):70

Fain P (2007) Still Young enough to be hungry. The Chronicle of Higher Education http://chronicle.com/weekly/v53/i32/32a03001 Accessed 4/11/07

Farrell JM (2009) Duncan: Community colleges important to restoring economy. Diverse Issues in Higher Education online http://www.diverseeducation.com/artman/publish/2009/03/09

Field K (2009a) Obama pick shows focus on training work force. The Chronicle of Higher Education http://chronicle.com/weekly/v55/i31/31 Accessed 4/6/09

Field K (2009b) Maine community-college dean is named to federal higher-education post. The Chronicle of Higher Education http://chronicle.com/article Accessed 4/8/09

Ford AR (2009) It's not just about racism, but ableism. Diverse Issues in Higher Education online http://www.diverseeducation.com/artman/publish/2009/04/02

Glenn D (2008) What Universities can do to graduate more minority PhD's. The Chronicle of Higher Education http://chronicle.com Accessed 12/4/08

Goodall D (2009) Bridging the gap from GED to community college student. Diverse Issues in Higher Education online http://www.diverseeducation.com/artman/publish/2009/03/05

Gose B (2008) Whatever happened to all those plans to hire more minority professors? The Chronicle of Higher Education http://chronicle.com/article/weekly/v55/i05/05b Accessed 9/26/08

Haines E (2009) Economic woes test historically Black colleges. Diverse Issues in Higher Education online http://www.diverseeducation.com/artman/publish/2009/02/16

Hannah DC (2009a) Are HBCUs still relevant? DiversityInc http://www.diversityinc.com/public Accessed 1/8/09

Hannah DC (2009b) Job discrimination hits record high: Is the economy to blame? DiversityInc http://www.diversityinc.com/public/5446 Accessed 3/13/09

Hayes D (2008) Community college mentoring programs help students stay engaged. Diverse Issues in Higher Education online http://www.diverseeducation.com/artman/publish/2008/07/24

Hennick C (2009) Creating a sustainable pipeline. Diverse Issues in Higher Education online http://www.diverseeducation.com/artman/publish/2009/02/18

Hoover E (2009) Community colleges anticipate boom in baby-boomer students. The Chronicle of Higher Education http://chronicle.com/daily/2009/04/15231 Accessed 4/6/09

Hunt J (2009) Changing the face of the legal profession. Diverse Issues in Higher Education online http://www.diverseeducation.com/artman/publish/2009/04/14

Jaschik S (2008a) Racial gaps in faculty job satisfaction Inside Higher Ed http://www.insidehighered.com/news Accessed 12/5/08

Jaschik S (2008b) Gates Foundation to spend big on community colleges. Inside Higher Ed http://www.insidehighered.com/news/2008/11/12

Jaschik S (2009a) Orientation for whites. Inside Higher Ed http://www.insidehighered.com.news/2009/04/13

Jaschik S (2009b) Rise in distance enrollment. Inside Higher Ed http://www.insidehighered.com/news/2009/04/06

Jaschik S (2009c) Morphing community colleges. Inside Higher Ed http://www.insidehighered.com/news/2009/04/07

Jaschik S (2009d) Community college surge. Inside Higher Ed http://www.insidehighered.com/new/2009/03/18

Jones SE (2006) Just What the PhD. Ordered. Spare Not Publishing, Birmingham, AL

Journal of Blacks in Higher Education (2009a) University of Wisconsin comes up short in meeting diversity goals http://www.jbhe.com Accessed 1/29/09

Journal of Blacks in Higher Education (2009b) MIT debuts new diversity website http://www.jbhe.com Accessed 2/12/09

Journal of Blacks in Higher Education (2009c) The persisting racial shortfall in tenure rates http://wwwjbhe.com Accessed 1/8/09

June AW (2008) On the road to tenure, Minority Professors report frustrations. The Chronicle of Higher Education http://chronicle.com/daily Accessed 12/5/08

Lederman (2007) A haven for minority scholars. Inside Higher Ed http://www.insidehighered.com/news/2007/10/30

Lederman D (2008) Who produces Black PhD's? Inside Higher Ed http://www.insidehighered.com/news/2008/09/02

Lederman D (2009) 2-year college leader is US education nominee. Inside Higher Ed http://www.insidehighered.com/news/2009/01/02

Limbach P (2003) Mentoring minority science students: Can a White Male really be an effective mentor? American Indian Graduate Center http://www.aigc.com/articles/mentoring-minority-students/2005/03/17

Matthews F (2009) Perspectives: a win in the diversity column. Diverse Issues in Higher Education online http://www.diverseeducation.com/artman.publish/2009/03/19

Metzler C (2008) Perspectives: Banning affinity groups shows lack of understanding. Diverse Issues in Higher Education online http://www.diverseeducation.com/artman/publish/2008/04/21

Metzler CJ (2009) The case against cultural standardization in tenure decisions. Diverse Issues in Higher Education online The Academy Speaks http://diverseeducation.wordpress.com/2009/04/06

Miranda E (2008) How one community college gave displaced factory workers the confidence to enroll. Diverse Issues in Higher Education online http://diverseeducation.com/artman/publish/2008/02/20

Moltz D (2009a) Diversifying Middle America. Inside Higher Ed http://www.insidehighered.com/news/2009/04/08

Moltz D (2009b) New approach to community college transfers. Inside Higher Ed http://www.insidehighered.com/news/2009/03/20

Moltz D (2009c) Who are community college trustees. Inside Higher Ed http://www.insidehighered.com/news/2009/04/06

Moltz D (2009d) Black men and remediation. Inside Higher Ed http://www.insidehighered.com/news/2009/03/31

Moltz D (2009e) Beyond "drill and kill." Inside Higher Ed http://www.insidehighered.com/news/2009/03/19

Monastersky R (2006) Brief Intervention Improves Achievement of students subject to negative stereotyping, Study finds. The Chronicle of Higher Education http://chronicle.com Accessed 9/1/06

Moore FL, Penn ML (2005) Finding your north. PotentSci, Emeryville, CA

National Academy of Sciences, National Academy of Engineering and Institute of Medicine (1997) Adviser, Teacher, Role Model, Friend – on being a mentor to students in science and engineering. National Academy Press

National Science Foundation (1997) Mentoring for the 21st Century Workforce

Nealy MJ (2008a) New study looks at retention in community colleges. Diverse Issues in Higher Education online http://www.diverseeducation.com/artman/publish/2008/03/24

Nealy MJ (2008b) New report highlights schools that make minority student success a priority. Diverse Issues in Higher Education online http://www.diverseeducatio.com/artman/publish/2008/04/21

Nealy MJ (2008c) Report: Black students among the most engaged at community colleges. Diverse Issues in Higher Education online http://www.diverseeducation.com/artman/publish/2008/11/17

Nealy MJ (2009a) University system of Maryland sets goals to close racial graduation gap. Diverse issues in Higher Education online http://www.diverseeducation.com/artman/publish/2009/03/19

Nealy MJ (2009b) Education Secretary Duncan: HBCUs as relevant today as ever. Diverse Issues in Higher Education online http://www.diverseeducation.com/artman/publish/2009/02/29

Nealy MJ (2009c) Do affinity groups create more racial tension on campus. Diverse Issues in Higher Education online http://www.diverseeducation.com/artman/publish/2009/03/10

Nealy MJ (2009d) Chronicling the lives of Native Americans on predominately white campuses. Diverse Issues in Higher Education online http://www.diverseeducation.com/artman/publish/2009/03/04

Nelson DJ (2007) A national analysis of minorities in science and engineering faculties at research universities. http://chem.ou.edu/~djn/diversity/Faculty_Tables_FY07/Final Report07.html

Parry M (2009) Sloan Foundation ends major grant program for online education. The Chronicle of Higher Education http://chronicle.com/daily/2009/04/06

Pluviose D (2008) More high-achieving students are choosing community colleges first. Diverse Issues in Higher Education online http://www.diverseeduation.com/artman/publish/2008/02/21

Pope J (2009a) Under a third of men at Black colleges earn degree in 6 years. USA Today 3/30/09

Pope J (2009b) Men struggling to finish at Black colleges. Diverse Issues in Higher Education online http://www.diverseeducation.com/artman/publish/2009/03/30

Porro J (2009) Reviving the great debaters tradition. Diverse issues in Higher Education online http://www.diverseeducation.com/artman/publish/2009/03/27

Redden E (2009) International applications up 4%. Inside Higher Ed http://www.insidehighered.com/news/2009/04/07

Reynolds CV (2008) Riding into college with your posse. Diverse Issues in Higher Education online http://www.diverseeducation.com/artman/publish/2008/02/27

Rivers LO (2009) HBCUs: Segregated but not segregating. Diverse Issues in Higher Education online http://www.diverseeducation.com/artman/publish/2009/03/09

Sander L (2008) At community colleges, a call to meet students at the front door. The Chronicle of Higher Education http://chronicle.com/weekly.v54/i29/29a02501 Accessed 2/28/08

Schmidt P (2007) Color and Money. Palgrave Macmillan

Schmidt P (2009) Doctoral universities pull ahead in competition for foreign graduate students. The Chronicle of Higher Education http://chronicle.com/daily/2009/04/15296

Schneider J (2009) Let's tackle education funding once and for all. USA Today 4/9/09

Selingo JJ (2009) Community-college leaders confront a challenge: enrollments are up but money isn't. The Chronicle of Higher Education http://chronicle.com/news/article Accessed 4/6/09

Sidanius J, Levin S, Van Laar C, Sears DO (2008) The Diversity Challenge: Social Identity and Intergroup Relations on the College Campus. The Russell Sage Foundation

Staff (2009) Minority-serving institutions called to serve as leaders in student retention. Diverse issues in Higher Education online http://www.diverseeducation.com/artman/publish/2009/04/08

Stuart R (2009) Surge in economic dropouts spurs UNCF into emergency mode. Diverse Issues in Higher Education online http://www.diverseeducation.com/artman/publish/2009/04/09

Sullivan P (2009) An open letter to ninth graders. Academe Online http://www.aaup.org/AAUP/pubsres/academe/2009 Accessed 1/20/09

Supiano B (2008) Taking courses as a group helps community college freshman succeed, study finds. The Chronicle of Higher Education http://chronicle.com/daily/2008/03/2004n Accessed 3/11/08

Tatum BD (1997) Why are all the Black kids sitting together in the cafeteria. Basic Books

Tirrell-Wysocki (2009) Recession sending more students to community colleges. Diverse Issues in Higher Education online http://www.diverseeducation.com/artman/publish/2009/02/10

Visconti L (2009) Can companies survive without a concern for diversity http://www.diversityinc.com/public/5106

Waiwaiole EN, Noonan-Terry CM (2008) Perspectives: Need to equip, prepare community college faculty has never been greater. Diverse issues in Higher Education online http://www.diverseeducation.com/artman/publish/2008/02/07

Warfield RD (2009) Perspectives: meeting the standards, not lowering them. Diverse Issues in Higher Education online http://diverseeducation.com/artman/publish/2009/04/13

Wilson R (2009) Critics challenge diversity language in Virginia Tech's tenure policy. The Chronicle of Higher Education http://chronicle.com/daily/2009 Accessed 3/26/09

Yue V (2008) Bonds to faculty help keep Latinos in STEM majors. Diverse Issues in Higher Education online http://www.diverseeducation.com/artman/publish/2008/09/10

Chapter 4
Precollege and College Preparation for Becoming a Scientist

Preparing for a career in science has to begin at a very early age, especially for the disadvantaged, in order for the individual to be competitive. In fact, for those interested in science careers, decisions are often made as early as the fourth grade, resulting in a significant loss of numbers of students, particularly ethnic minorities, for future science careers at a very early age. As such, it is critical to not only expose young people very early to positive influences of science but also to engage them in a way that will help them to continue to stay on track and actually build up their experiences so as to make them most competitive in the field. Although the "things" that they need to do are easily identifiable, the key component is having mentors there to assist them, advise them and generally help to guide them in taking full advantage of the opportunities, and in doing so, strengthening their credentials so that they will be successful. Interestingly, many of the things that students need to do in K-12 are similar to those that they need to do as a college student, obviously just at a different level, which is why the earlier they start, the better prepared and more competitive they will be.

General Considerations

The problems associated with the leaky pipeline of minority scientists definitely exist at all levels from precollege to transitioning into careers. However, the total numbers from the under represented ethnic groups in the sciences at the precollege level is significantly high, as would be expected by the changing demographics of our society. Still, unfortunately, of those numbers, the proportion of those students interested in science is greatly diminished, because of a number of reasons, perhaps the most important of which is the exposure, both lack of positive exposure and too much negative exposure. In fact, exposing K-12 children to the sciences in a most positive way followed by effective mentoring is probably the most important aspect of addressing this problem of under representation, since many students are lost to the area of science early in their education experiences, e.g., fourth grade, for all the wrong reasons. These reasons include the negative stereotype of scientists as

T. Landefeld, *Mentoring and Diversity,* Mentoring in Academia and Industry 4,
DOI 10.1007/978-1-4419-0778-3_4, © Springer Science+Business Media, LLC 2009

portrayed by the entertainment industry, e.g., *The Nutty Professor* and *Honey, I Shrank the Kids*, i.e., the notion that scientists are isolated in their work, a lack of knowledge as to what a scientist really does, delayed gratification, the fallacy that all scientists are geeks/nerds and of course, the notion that mathematics and science is just "too hard." In addition, the under representation of minorities in the sciences, e.g., professors, professionals, teachers, significantly adds to this problem with young children, especially those from under served groups, because they do not really get the chance to see many scientists that actually "look like them." Additionally, due to the current shortage of science-trained science teachers, particularly in schools serving students from these ethnic groups, they really do not see the type of enthusiasm for science that is needed for students in that age group. Addressing the last point requires a major reemphasis in our teacher training (and incentives) whereas some of the other reasons can be ameliorated by mentoring efforts of faculty and students, i.e., college, graduate, professional. In fact, minority undergraduate and graduate/professional students can play a pivotal role in exposing the younger students to positive images of scientists by taking time to visit their K-12 schools, talking to relatives at family gatherings and in general being involved with young people. And do recognize that this type of commitment has to be real and not veiled in some program that does not work now or possibly in the future. Such an example could be the commitment to K-12 education particularly in the science area as put forth by Dr. Charles Reed, Chancellor of the CSU system, which has over 400,000 students enrolled in *Diverse Issues in Higher Education* (3/09). Despite what appeared to be a major commitment to this area by the CSU Administration, other demonstrations of the lack of support for faculty and staff issues as well as supporting administrative raises rather than increased enrollment of students portrays a very different image to this commitment, especially since it is the faculty that would be directly responsible for the education of these future teachers. "Walking the walk" rather than just "talking the talk" is an absolute must in the success of such efforts.

Specific Tips for Preparation

Interestingly, the things that all students, and particularly those from under represented groups, both in K-12 and college, can do to better prepare for an education and career in science are actually quite similar and that is why exposing students at an early stage to positive images and influences about science serves as a very important first step in the process. Then, as the student progresses through their educational development, the reemphasis of the positive images and influences remain as a constant reminder and incentive for them. Despite this, the "science market" is an extremely competitive one, and exposure of the students is not nearly enough. They *must* not only be made aware of what they need to do to be successful but also actually receive advice, i.e., mentoring, continually along the way by individuals who have the appropriate expertise and experience on how best to do these so as to reach their goal. In some cases, these are very straightforward, such as taking the appropriate courses and, of course, doing very well academically in them, whereas others are aspects that students who are not interested in pursuing advanced degrees and professional careers may not even attempt, e.g., taking standardized tests or writing a personal statement. It is in fact these aspects that will eventually separate the successful ones from the unsuccessful ones, which is even more of a reason for the student to be mentored as strongly as possible. As for the relative importance of each of these efforts in determining success, there is no real numeric value that can be applied, e.g., #1 vs. #2 but rather it is a culmination of all of these taken together that usually dictates success. As such, the following are listed in no particular order (even though they are assigned a number); however, there are some very important "mentoring tips" associated with each of them, i.e., just identifying these will not serve as enough information for most students to be highly competitive without tips on optimizing that particular aspect with regard to use in future applications. The fact that the "tips" are more extensive than others relates to the fact that there is considerable latitude in the preparation and formulation of that particular feature, e.g., personal statement.

1. *Seek out, identify, and "nurture" the relationships with mentors* – Identification of mentors represents an even more important decision than choosing your career and/or the best way to prepare for that career, since mentoring will play a critical role in whatever career path that you decide. Importantly, these individuals not only are most valuable at the time that they are initially identified (and subsequently nurtured) but also throughout your life. In fact, Dr. Frank Talamantes, a trusted colleague, dear friend, and valued mentor once said "mentors are like tattoos, once you have them you have them forever." I took the liberty of adding "if you lose one, it hurts like heck!" Since choosing a mentor is so important, it must not be done lightly, even though there may be certain instances where it may happen almost by chance, e.g., being at the right place at the right time. The point is that once you and that individual recognize your similar interests, likes/dislikes, common areas, etc., the onus is to then nurture that relationship to the fullest degree. This is not difficult but does take an effort that requires attention, focus, and specific interactions, on the part of both the mentor and the mentee. However, since it is likely that the mentor has other mentees, it is more difficult for the mentor to expend the same amount of effort and energy in nurturing the relationship. Thus, the effort must come primarily from the mentee. From there, it is critical to continually update the mentor with your progress, your accomplishments, your failures, i.e., everything that "comprises you" and your efforts to achieve your goals. You might say "this sounds like a developing friendship," which is exactly what it is, especially when you consider that it should last the better part of a lifetime (even though there is some discussion and disagreement whether of not friendship is the best word, see above). Granted, the actual relationship will change as both you and the mentor continue along your paths but that too is to be expected. For example, for a young man who I mentored to get into medical school, assisted while he was a medical student, mentored as he got

into his profession, we now actually will socialize together in certain settings but the mentor–mentee relationship always remains. Importantly, there have definitely also been times during this relationship where I have learned from him, i.e., I have been the mentee. Plus I know that he has served many times in the capacity of a mentor for others and in fact have asked me at times for help in mentoring these young people. Also, the role of the mentor is to identify networking opportunities based on their experience and I will share examples of this later. So do you choose someone that looks like you, e.g., same ethnicity, same gender? (Petchauer 2008 "But Professor? You're not white, you're German, right?"; Limbach 2003 "Mentoring minority science students: can a white male really be an effective mentor?"). How about their involvement in the same or similar discipline, e.g., medicine, research? How about similar backgrounds, e.g., urban vs. rural upbringing? Well, the answer is that any/all of these may be important or maybe, but unlikely, none of them. Granted, similarities of some kind are normally what bring us together but this is not always the case and these are not always immediately visible. What truly becomes most apparent and important is "who they are and what they are about" – something one learns by spending time together. Once this is established, then the relationship can be developed to the degree that you desire. Like any relationship, establishing it is only the first step; by not following up and building upon it, you lose the whole purpose. To that point, remember, as mentioned, the mentor most likely mentors others so again the onus is upon the mentee to foster and maintain the relationship. Finally, as also mentioned earlier, mentors should be lifelong relationships, to the degree possible. Granted, once the mentee (or the mentor) moves on from the setting where the relationship was initially established, e.g., college, it is difficult to "maintain," but not impossible, similar to when a friend moves away. One needs to make every effort possible to maintain contact and continue to update their mentor for as long as possible, especially when one realizes truly what a small world it is. For example, the medical student that I mentioned earlier, upon graduation went to the East Coast for a residency, whereas I moved to California with a new position. Fortunately, we did stay in touch, and then he took a position not only on the West Coast, but in the same area of LA where I worked! We were never "disconnected" but upon his relocation to the West Coast, we definitely reconnected on a more regular basis. He has since relocated back to the East Coast, but we correspond regularly and if he is on the West Coast, or me on the East Coast, we make ever effort to get together. Moreover, it is critical to continue to develop mentors in each of the major settings where you are developing your goals as each one can contribute important information relative to the different stages of your career development, as really, over the course of a lifetime, the number of mentors that you have will probably be similar to the number or "true friends" that you have.

2. *Securing letters of recommendation* – All applications require letters of recommendation, very often three in total. As such, it is advisable to secure four as a safety measure since one never knows what may happen to one of the individuals including them just being negligent in submitting the LOR. If only two LORs are

received by the Admissions office, when three are required, one's chances of being admitted fall precipitously. The student often wonders who should write these letters. This choice is of paramount importance as letters of recommendation can make or break a person's application to the next level as these letters must speak effectively and confidently about the student as an individual, i.e., these letters must be of substance, containing specific and pertinent information. Therefore, you must first select these persons very carefully and, even more importantly, nurture the relationship with these people over time (see #1 above) so they can speak confidently and comprehensively about your qualities in their letter of recommendation. The best advice for who should write these letters is: "people who know the student best." As such, most often these individuals are mentors for the student, since as indicated earlier, mentors must know students very well, both professionally and personally. For this reason, the letter writers do not necessarily have to be in the science discipline, although at least one letter should come from a science professor/health advisor. Others that may write the letter could be a supervisor, a professor from another discipline, the head of a volunteer organization – anyone who can speak to the student's qualities and potential. Individuals who should *not* provide a letter of recommendation include relatives and friends of the family, as these are usually judged to be biased. Additionally, even though letters from people with impressive titles are helpful, if the person does not know the student well and therefore cannot write in detail, such a letter which would read like general character reference would not do any good and may actually be harmful. Some other helpful suggestions in securing these letters of recommendation include: (a) Early in your studies, i.e., fresh person or sophomore, identify three (and if possible four) people that you believe can write you strong letters of recommendation (LORs). Most importantly, they have to know you *well* as a person! Meet with the potential reference and *ask* if she/he is willing to write good reference letters for you. Let the person know what the letters are for and approximately how many are needed before asking for a commitment. To ensure "good letters," *nurture* the relationship with the individual by providing them with information about yourself (i.e., personal statement, recent achievements) on a regular and updated basis. Very importantly, to know that it is a good LOR, ask to see the letter, as this is the *only* way that you will know for sure. There is nothing that says the letter has to be confidential. Letters that are based on little contact, containing any innuendos or any negative statements can *kill* your application. If you nurture your relationship with these individuals, obtaining a copy should present no problem. If the person agrees to be a reference, then proceed with the next steps. (b) Provide a folder containing the deadline dates, a copy of a personal statement and resume, plus any evaluations forms that might be required to be completed. (c) Be sure to allow them sufficient time to write the letter whether it be requesting the LOR, for the first time or after you have had additional accomplishments, *always* allow adequate time for the recommender to prepare it. *Do not give them the request the day before it is due or even with a short time deadline, such as a week or less.* (d) If the person says "no" or is hesitant about writing the letter of recommendation, identify someone else rather than try to convince him/her! (e) Follow up at least a week before the due date. (f) Keep your references updated as to your progress on a regular basis, e.g., if you received recent recognition or are accepted for an interview. Remember the more they know the stronger the letter! (g) Write a thank you note and ultimately let all references know the outcome of the application. (h) Continue to maintain contact with these individuals even after you leave the setting, e.g., graduation, for a period of at least 1–2 years, updating them on a regular basis. As one additional note, if the person asks you to write your LOR for them, you should be very concerned as (1) that is *not* your responsibility; and (2) it suggests that writing a LOR for you is not as important to them as it should be. In response, indicate to them that you are not comfortable doing that and if they still insist you must then decide to either do it or ask someone else for the LOR.

3. *Strive for the best grades possible, especially in the science courses* – Grades are obviously a standard that can be used across academic institutions and programs, regardless of claims of possible "grade inflation" and/or quality of the institution. For that reason, it is essential to perform as well as possible in one's curriculum and for students interested in a science career, strong performances in the science courses is crucial. In fact, being able to enroll, and do well, in advanced science courses demonstrates even stronger on one's potential as a scientist. No, one does not have to accrue straight A's, in either science or non-science courses; however, poor performance in key science courses will definitely raise serious questions, if not concerns, about your ability to perform and succeed at the next level. Dealing with the occasional poor grade or even a semester can be discussed as part of one's personal statement (see later). With that being said, perhaps the most important thing that one must consider as one performs academically is why one does poorly. Sometimes, on the one hand, it may be a true lack of interest, e.g., studying premed because of your parents. On the other hand, it may be due to the lack of preparation, e.g., high school. If this is the case, then this can often be remedied through certain types of remediation. Finally it may be that one just cannot comprehend the material. Although this may be a combination of the previously mentioned factors, one needs to seek assistance in determining that before choosing something else. The focus on doing well, specifically in the science courses, becomes even more important as one progresses through their academic experience, e.g., performing well in undergraduate organic chemistry and biochemistry courses is evaluated very critically by admissions committees for the health professional schools. As the courses become more closely related to those in the graduate/professional schools, it becomes equally important to perform well academically. Although many majors only require a C in the courses within their major, students must perform at a much higher level to compete successfully for admission into post-baccalaureate programs. It is *not* necessary to maintain a straight "A (4.0)" record; however, numerous "C" grades in courses that are judged to be important in further education programs are often viewed very negatively. Even more damaging is several "withdrawals" (W's) as these send a message to admissions

committees that the student cannot persevere in classes that require effort. Thus, it is imperative that one works hard to maintain a GPA that truly represents their abilities in academic courses that are prerequisites for further degrees. If poor time management is not the contributor to the lower GPA, then the student must look hard at why the grades are not better and ultimately if this major is perhaps not really the one for them, especially as one progresses through their academic program, e.g., very often grades can suffer upon entering a new environment, i.e., freshman and as such this is often seen as acceptable by admissions committees. However, poor performance later in the student's matriculation, and in particular continued poor performance, is very often a reason for rejection of an application. One aspect of taking courses that admission committees look at carefully is the number of science courses taken at one time, since this would be the type of schedule that the student would take in the graduate/professional program. Many undergraduates will take one or two science courses along with other nonscience courses, e.g., general education (GE) courses. As such, the difficulty of the science courses can often be "buffered" for the student, making it difficult for the admissions committee to get a true assessment of how the student might do when taking a number of science courses at one time such as will be the case in the graduate/professional program. This is one of the big advantages and strengths of Post Baccalaureate Programs as the participants usually only take sciences courses, making it easier for the committee to assess their performance at a level more analogous to graduate and professional programs.

4. *Take advanced science courses* – As mentioned earlier, this is not absolutely necessary but can be viewed quite positively for those who have succeeded in doing well in these courses, primarily because it demonstrates the ability to work in a concentrated and focused manner much like is the case in graduate and professional school. Also, it is hoped that students take the courses due to their interest and not just for the "show." For the high school student, taking Advanced Placement courses very often helps in their admissions into college and in many cases reduces the classes that they have to take. At the very least, these courses provide the student with excellent preparation toward their major fields of study. However, it is important to note that there is seen a "racial divide" in the results of students taking advanced placement high school courses (Elfman (2009) "AP results improve for other minority students, though Blacks still lag behind"). These results that are presented in that article come directly from the "AP report to the nation" and shows that Black students are less likely to pass the examinations or even take the classes than other minorities and of course nonminorities. It is suggested that this is due to inadequate preparation for the courses. The effects of this can be seen later on as normally this saves the students money as they can earn college credits while in high school thus allowing them to graduate from college sooner. As a result of these findings, the College Board, who monitors this AP program, recommends that states fund efforts designed to assist students in preparing for AP examinations, such as teacher workshops. Those states that have done this have seen the gap close between white students and minorities, e.g., Hispanic in states such as Texas, Florida, and California. Another

finding that can be most beneficial to graduation rates of under represented minority students from high school is the "early college program" (Moltz 2009 "Promoting early college"). Although there have been several such programs described, one of the innovative ones is the Early College High School Initiative, which is coordinated by Jobs for the Future and funded by the Bill and Melinda Gates Foundation. This program recognizes more than 250 "early college high schools" in 24 states and the District of Columbia, which encourages over 100,000 students to not only graduate with a high school diploma but also with an associate degree or 2-year's worth of college credit. Other such programs include the Bard High School Early College, which was the model that Gates' funded program was based on as well as the Gateway to college. Back to considerations at the college level, although a number of upper division courses, particularly in the sciences, can prove beneficial once the students enters a graduate or health professional program, probably none is more important than biochemistry as it is truly a discipline that crosses many others within the field. Beyond that, choices of courses should be based also on interest and relevance to the path and ultimate career that is chosen

5. *Get involved in research* – The exposure to research during one's academic training is extremely valuable *regardless* of what career choice is made as research provides the individual the ability to think critically and analytically in a way that no course can ever do. In fact, it is difficult to explain what research really does in preparing an individual for a career in science; one just has to experience it. At the same time, although what one learns for doing research is invaluable, the person may also decide that this career path is "not for them." Regardless, they are better prepared to succeed in their chosen profession. Thus, will a physician who conducted research at some time during their training be a better physician? It is hard to say "definitively yes"; however, many truly believe that the physician will definitely look at certain cases in a more comprehensive and analytical manner than those who did not have any exposure to research. The research experience can be varied in that it can be basic, clinical, or a mixture of the two, actually that does not matter as much as the fact that the student is involved in a research project. Should this experience be during high school, undergrad? Should it be during the school year? During a summer? There is no right answer other than it must be an experience in which they are totally immersed so that they are truly part of the project, i.e., not just cleaning glassware. They must know the project goals, the reasons for the experiments, be able to interpret data and make conclusions. At the same time, this does not mean that one *has* to have a publication from their experience whether it be a paper or an abstract, although if that is the case, it certainly strengthens that research experience. At this point in time, for anyone wanting to enter a PhD graduate program, doing research before entering graduate school is not an option, it is essentially a requirement. There is no doubt about it – *the PhD is a research degree!* As such, being part of a published paper, abstract, presentation is becoming more and more necessary to be competitive for a position in a graduate program at a research institution. Because of this, some students come to graduate/professional programs with as

much as 6 years of research, i.e., 2 years in high school and 4 years in college. Considering the competitiveness of applicants, one needs to assess their involvement accordingly. It should be noted that those 6 years may not be six "full" years; in fact it could be a summer program for six different years. Regardless, the exposure and experience will add significantly to the strength of the application. Also participation in summer programs external to the student's institution is invaluable as it not only demonstrated a willingness to explore a new setting but also serves as a basis for an application to that institution for their graduate/professional tenure, if their experience was most positive. Finally, it is important to recognize that research now tackles bigger and bigger challenges, e.g., global warming, AIDS, which in turn requires more and more collaborative and interdisciplinary efforts. As such, choosing a research program, especially early in one's career, e.g., high school or college, should not be made solely on an area where one believes they want to "spend their life." In fact a variety of research settings will add most positively to a student's research experience as they continue on into their other educational pursuits.

6. *Research and visit college programs* – As one might imagine, programs can be very individualized, i.e., certain students will be attracted to certain programs. The onus is for you to identify what you are seeking from a program both for a baccalaureate degree as well as a degree following the bachelors. Thus, once you have identified your preferences, then you research those schools/programs that meet your "requirements," e.g., what you are seeking can be as simple as location, e.g., staying near home or avoiding cold weather to as complicated as the type of degree that is offered. As mentioned in an earlier chapter, consideration of a community college as part of your undergraduate experience can be a critical component in your decision making. If that is the case, then those things to consider are those that work best for you, just as in other decisions that you make. The path that students take to graduation and achieving their goals can be quite "jagged" (Associated Press 2008, "Some students take jagged path to graduation"). Unquestionably, visits to undergraduate institutions should include parents since in many cases they are still intricately involved in the education process, often including substantial financial support. For the advanced degree, e.g., graduate PhD programs, there are additional considerations. For example, many settings now offer an umbrella degree in the biomedical sciences, as opposed to one specifically in pharmacology or physiology. As such, it is especially important to research these programs well to see a "fit" between your interests and the areas within the program, especially since not only the specific discipline is important but also the type of research that is being conducted. In the medical arena, there are MD granting schools (allopathic), DO granting institutions (osteopathic) as well as other medicine programs, e.g., Naturopathic physician, a doctor of podiatric medicine (DPM). Students need to know the differences in the various programs, which can be determined both by advise/counsel from mentors as well as by the student from the Internet and other published materials, e.g., http://www.explorehealthcareers.org. Probably most importantly is to see "what works

best for them," which is also often determined through mentoring, development of personal statement, direct interactions, etc. There are a number of combined degree programs as well that can meet the need of individual students, both from the standpoint of matriculation and also their ultimate career choice. These include the MD/PhD, the MD/MPH, the DDS/PhD, etc. Oft times, these will be specifically funded programs such as the Medical Scientist Training Program (MSTP) for the MD/PhD or simply a collaborative effort by the schools that grant the respective degrees. Finally, in recent years, there has been a movement to a professional Master's degree; in fact a number of these programs have been developed. Since many graduate programs in the past have awarded master's degrees only if the person could not complete the PhD program, one of the issues that these programs have to deal with is "raising the profile" of the Master's degree. Regardless, these degrees represent an alternative for some students and will be discussed more in the chapter on graduate vs. professional school. Unquestionably, the more that the student knows through whatever resources, the better it will be in deciding what best "fits." It is very important to realize that the costs for visiting potential sites to matriculate, i.e., interviewing, are quite disparate, e.g., most graduate programs, at least in the biomedical sciences will "foot the bill for the visit" whereas in most cases interviews at health professional programs/schools pay nothing.

7. *Seek out possible funding, e.g., minority-targeted* – As mentioned earlier, there were a number of efforts directed toward effectively addressing the under representation of minorities in the sciences, one of the most major of which was summer research opportunities at major research institutions across the country. As more attacks on these opportunities for minority students occurred, those programs specifically targeting under represented groups began to diminish. Although many of the programs for summer research still exist, many of them are open to all students now, thereby decreasing the specific opportunities for minorities. Regardless, students need to investigate these opportunities at universities, foundations, government agencies, and other institutions across the nation, regardless of their focus. To that end, there are a number of websites that provide listings of such programs nationwide. These are not necessarily exhaustive but they are quite comprehensive and when taken in total they provide quite an array of opportunities. In addition, lists are often provided in conventional publications/periodicals, e.g., *Ebony* publishes an extensive list every year in one of their issues. In all cases and in particular in the case of *Ebony*, most of these opportunities are very specialized, so it is important to look carefully for any and all of these that "fit" for each individual student. As an example, I wrote a significant number of recommendations for a particular medical student when I was at Michigan, due to the fact that she was constantly identifying sources of funding for which she was uniquely qualified (and this was even before the Internet was so popular and prevalent!). Regardless of minority vs. nonminority, seeking external funding is very important, not only for the obvious reason of helping to defer expenses but also because most funding is award-based and as such can be an important part of one's "honors" on the resume.

8. *Develop a strong personal statement* – The personal statement represents one of the most important components of an individual's application to any program as it conveys to the selection group the strengths of the applicant, beyond the GPA, test scores, etc. In fact, from my extensive experience and involvement with admissions at the medical and graduate school levels, I saw the personal statement as the most important component of the student's application. As it turns out, unfortunately, most of our educational systems do not provide students with complete or sometimes even proper information for preparing a strong personal statement. Why is this the case when throughout high school and college there are career counseling offices? An easy answer to this question is that there are just not enough counselors (mentors) for the number of students, let alone the fact that not enough of the counselors are trained in many of the specific areas of science careers. For that reason, many students have no idea how to write such a statement; in fact, many do not even know what one is. A mentor is needed to deal with both these points. So what is a personal statement? To cite the obvious, a personal statement is a description of who you are as a person and how those personal traits/characteristics will contribute to the program/institution that is responsible for preparing you for your career as well as how these traits will make you a positive influence in the career that you have chosen. Mostly, students often have the problem of getting started. So, to get started, one should list their top 2–3 qualities as a person, i.e., what makes them who they are. They should then ask a friend what she/he sees as their strongest, most positive personality traits. Do they match? Similarly, they should ask a family member the same question. Once there is agreement or at least a consensus of those top traits (that of course the student agrees with), they then have the basis for their personal statement, i.e., this is what should come through when someone reads their statement. This is a personal statement so it has to reflect who the person is. At that point, it is important to then create a story that incorporates those traits into their plans for a career, including why they chose it, where and how they have been preparing for it, and, of course, the immediate step in this process, i.e., the educational background required to enter into the profession. Making this statement a well organized, smooth flowing story is critically important as admissions members read many of these and just like a book, if the initial reading is not interesting, novel, or exciting, chances are that the member will move onto another one. Moreover, to make this story read well, the facts/information must be tied together by transitioning one idea/phase to another. Very often, initial drafts, and occasionally, final copies of these statements are "piece mealed" and thus, although the information is there, it does not read well and therefore often will not hold one's interest. This is why initial drafts will often be much longer than one page so that more final drafts can be shortened and revised. Once again, input from various individuals is invaluable as it allows for different perspectives. This essay is a window demonstrating how the student thinks since how one writes is a reflection of how well one thinks. General tips include the following:

   - Before even writing the statement, start by listing your top 2–3 qualities as a person, as these should be in your theme throughout the statement. Once you have done this, ask others including friends who know you well and also family to do the same and see how closely the lists parallel one another.

- The first sentence in the first paragraph needs to catch the Admissions Committee's interest. Give it some punch! Make it an "Attention getter!" Often a quote or a question will serve as an excellent "attention getter."
- If you have trouble starting, take a historical approach, e.g., I was young, I grew up, etc. However, if you can write creatively, by all means "create," rather than conform.
- Look at the first sentence of each paragraph (the topic sentence) and determine whether it is memorable or generic. Go for memorable. Also make sure that you "transition" from one paragraph to another. Just listing facts, even if they are chronologically relevant, is usually boring.
- Avoid starting sentences with long phrases, such as "Because I have always been interested in science since I was very young…" Rearrange the sentence to avoid these phrases.
- Read each sentence to see if all other applicants could have said the same thing. *Be unique*. What is there specifically about you that differentiates you from the rest of the applicants? The tone of your sentences should go up, leading the reader up to an important point.
- Show motion in the story you are telling. Tell about the *process*, rather than just the results.
- Be specific wherever possible, but do not go overboard, e.g., do not list methodologies in the lab (unless research specifics are requested).
- Tell enough about yourself so they can conclude your potential.
- Do not start a lot of sentences, in a row, with "I" (even though it is a "personal statement"). In fact, do not overuse "I."
- Stories are a wonderful way to make a point and also to keep the Committee's interest. In fact, you might even ask someone close to you such as a parent, a sibling, or a friend about a "best story" about you.
- Avoid the use of soft words/phrases. Use hard words/phrases instead. For example, "I think" is soft whereas "I believe" is hard. Similarly, a soft statement would be "I would like to help people" whereas a hard statement would be "I will then be in a position to apply my knowledge." Be assertive but not arrogant.
- Explain, but do not apologize. Everything is a success story in process, e.g., a bad semester should be explained, briefly. Remember that committees look for negatives. When doing this always lead with a positive statement, followed then by the "less positive" note.
- If possible, use humor in a positive manner as the statement should reflect enjoyment.
- Informal is okay but avoid slang.
- Definitely show your personality and your human side.
- Highlight your interests as they apply, e.g., as a physician, if you are interested in Family Medicine, by all means, say this. If you plan to work in an underserved or rural area, say so. Specific is better than global, *if* that truly reflects you.

- When you have finished the essay, read through each sentence to see if there are any words you do not need. You have a limited amount of space and even more importantly, you want to write concisely to convince the Admissions Committee that you think clearly.
- Be careful not to step on any toes. For example, you do not want to say "No one but a physician can truly experience the ultimate satisfaction in giving." First, this is not true and second, that type of statement will most likely offend someone, possibly on the admissions committee!

9. *Prepare for standardized tests* – Personally, I define these examinations as "necessary evils" in that schools require them in their assessment of applicants despite the fact that evidence for their validity in truly assessing a person's potential for succeeding in the program, let alone the profession, is not strongly supported/warranted. Perhaps a major reason for that is that there are no "standardized students" so who are the tests really assessing, especially when throughout the years they have continued to demonstrate cultural biases. Still, since these are required, students must do whatever necessary to best prepare for these examinations. Of course, there are preparatory courses, many of which are costly, not always readily available and not always effective. Most importantly, the student needs to know themselves best, particularly how to prepare for such examinations. I have seen excellent students not do well on standardized examinations even after taking preparatory courses. Interestingly, those same students have done outstandingly well in the programs for which the examinations were supposedly testing, including standardized examinations later in the program. Probably most importantly, especially for the health professional examinations such as MCAT, PCAT, DAT, etc., the best rule is to not take the examination until you are as ready as possible. This is extremely important as the examination scores stay with you regardless of how many times you take the examination as selection/admissions committees always look at *all* of your scores. For that reason, retaking the examination should be considered very carefully, i.e., identify what was not done the first time in preparing for the examination and improve on that before the next time. Data have shown that retakes of some of these examinations very often only result in a minimal increase in the score, e.g., 1–2 points, which the Admission Committees do not like. After all you took the examination twice! Besides preparatory courses and studying, generally, reading several newspapers daily as well as staying up to date with current events is an excellent way to help prepare, e.g., expanding vocabulary and keeping up on current events. Because of the nature of graduate studies vs. health professional schools, the GRE results are not as critical as an acceptance criterion as the MCAT, PCAT, etc. Students still need to prepare, and one of the best ways is to take the GRE practice tests on the ETS site; however, their GRE scores will not factor in as importantly to their acceptance as their research experience and letters of recommendation. As just a commentary regarding standardized test scores and their reliability is predicting success later on, California residents are not required to take the PCAT examination to get into California Pharmacy Schools. Obviously a number of excellent students graduate with PharmD's from these schools every year without having taken the PCAT!

Of course there are also many other scenarios where students have not done well on the standardized test required by their professional school but have gone on to perform very well in their classes in the curriculum as well as on the standardized tests that they take later in their curriculum, e.g., LSME Boards Part I and II for medical school. As stated earlier, students need to prepare for these examinations in whatever manner best suits them to be able to perform as well as possible on them. On a positive note, a recent report from the Josiah Macy, Jr. Foundation in looking at the current state of medical education, including the anticipated shortage of physicians over the next few years, made a number of cogent recommendations (Mangan 2009 "Medical schools should re-examine admissions and training methods, experts say"). One of the major of these recommendations was that the admissions process be reevaluated with much less of an emphasis on MCAT scores, thereby attracting a broader range of potential students, particularly those from under represented ethnic groups. As a result, a much more representative physician workforce could emerge. What about standardized tests and high school students? This issue of standardized tests and their "applicability" especially when considering minorities also extends to high school students and the SAT. *The Journal of Blacks in Higher Education (2/12/09)* reported that the racial scoring gaps, i.e., the difference between scores for whites versus Blacks, have increased on 10 of the 11 most widely taken SAT subject tests over the past 9 years, with only exception being physics. What is especially relevant to students with aspirations for careers in the biomedical sciences, the one SAT II test that has shown the largest increase in a scoring gap has been biology, which went from a gap of 67 points in 1999 to 101 points. Similarly, the racial scoring gaps have increased also in the areas of mathematics and chemistry. As for possible reasons, the article discussed the fact that a broader cross-section of Black students are now taking the SAT II tests, when compared with 9 years ago, and if that trend continues, it is likely that there will be an even further widening of the racial scoring gap. A potential result of this is of course minority students turning away from science careers and further negatively affecting the under representation. And most importantly, do SAT scores provide college admissions committees anything more than how well a student takes the test? In an article in Inside Higher Education online (3/26/09) by S. Jaschik, the issue of dropping the SAT, as opposed to some efforts to just make it optional (Jaschik 2006 "Momentum for going SAT-optional"), is discussed. Based on a study, ending the requirement of SAT or ACT for admission would lead to demonstrable gains in the percentages of Blacks, Latinos, and students who are working class and economically disadvantaged. The findings also further confirmed the fact that reliance on SAT for admissions favors applicants who are white and/or wealthier than other applicants.

10. *Be active in a number of activities that reflect "who you are"* – This is related directly to the personal statement in that whoever the student is should not just be "stated" but also reflected in their activities. The person who states "I always wanted to be a doctor" but does not demonstrate any efforts in their experience that would serve to define that is essentially compromising their integrity and,

as a result, committing professional suicide. This actually should be easy since you have a tendency to participate in activities that you like. At the same time, just do not list activities to "pad" the list. Make sure that these are relevant to your story. Involvement in such activities is valuable not only for the reasons already mentioned but also because it serves as part of the "networking connection," which as discussed later is critical in your development and success. Very often an activity may not appear to have direct relevance, e.g., captain of the cheerleading squad. However, the teamwork and leadership aspect of such an activity may compliment exactly what the student is exhibiting in their personal statement.

11. *Apply to several colleges/universities* – Many students often think that it is negative if they get accepted someplace and turn it down. As a result, they often limit the number of places where they apply. This belief is not true; in fact, programs expect this since they are in a very competitive marketplace. Let the program make the decision regarding you and then you can decide on the program. At the same time, the "shotgun approach" for applying does not work. It just costs you more money. All programs have acceptance criteria and many of these are very similar across institutions. Thus, not getting into a program most likely means not getting into a number of others. I had a student referred to me who had applied to 75 allopathic medical schools. He would have been much wiser to select more carefully the schools for "his fit" (including his strengths and weaknesses). Similarly, I had a student who was interested in primary care tell me that she was applying to the University of Michigan Medical School (UMMS) but not Michigan State Medical School. Not that UMMS could not provide her with an excellent education; however, I alerted her to the fact that: (1) Michigan was a tertiary care hospital, (2) Michigan State University specialized in primary care, and (3) MSU actually also had an osteopathic school of medicine. It is critical that one is best informed about the choices that have to be made. Once again this can be best accomplished through mentors *and* information materials from the appropriate resources. As a general guideline, for example with medical school applications, choose 10–20 schools to apply to, always with personal reasons for choosing them.
12. *Networking* – What is networking? The process of interacting with others to develop relationships, reach a goal or gather information. For this reason, networking is critical to all future educational and career goals. Life in general is all about "hook-ups" and progress/success in academia is no different. Some tips include (1) have a general prepared "blurb" about one's self and what one wants to do, not as a script that is memorized but rather used as a basis of introduction, e.g., a 3–4 sentence introduction about one's self and plans. One *must* be able to talk about one's self!; (2) show initiative by introductions; (3) have business cards available for meeting someone new; (4) during conversations, asking intelligent questions, responding succinctly but completely, and being respectful; (5) collecting the person's business card and afterwards while still fresh, noting anything special about the person/conversation for future reference; and (6) following up with the person through email. Although the value

of networking is immeasurable, it is sometimes not truly appreciated until that instance arises where someone connects from the past but one who was not expected necessarily to have an impact, especially at a given point in time. In many instances, the connections are many and often indirect. An excellent example is a young student that I met at a national minority conference over 15 years ago. She came to me asking for assistance in her eventual goal of becoming a physician (at the time she was an undergraduate in a minority-supported research program).From my experience working with such students, it was immediately obvious to me that she not only had the potential and dedication to accomplish what she wanted but also that she would use the network to help her achieve that. Throughout her undergraduate years, I continued to mentor her even though she attended a university different than mine, in fact our institutions were located almost across the country from one another. As such, we personally saw each other only through joint conferences. Still, through phone calls, emails, etc., I mentored her in applying for medical school, which was her next step. Because of a number of factors, her application was not as strong as we would have liked so I "connected" her with a program in Chicago with which I had previous experience, specifically with the Director and others, a program that focused primarily on students like her. She was able to get into that program and did well enough to successfully get into medical school. While in medical school, she faced additional obstacles similar to ones that she faced before, e.g., standardized tests, and, as such, I hooked her up with another colleague who specialized in test taking, especially for under represented groups, which are disproportionately affected by these tests. With that connection, she was able to move on into her clerkships years but again came across difficulties associated with the next set of standardized tests. Once again, she utilized the person from before whom she met through my networking. As the time came for placement in a residency program, difficulties were still being experienced due to the standardized test issues from before. Again the person who had helped with the standardized tests had connections with a residency director at a hospital in the city where she hoped to match. Moreover, a daughter of a close personal friend of mine (we had interacted socially not professionally for many years) had matched in a residency program in which my mentee was interested and in the same city. I hooked them up and through that connection she touched base with the director of that residency program who provided her with insights into getting into such a program and, in fact, even provided her with a mock interview. Thus, over a long period of time while this student progressed through numerous obstacles and roadblocks, I continued to mentor her and provide her with networking contacts that assisted her greatly in achieving at each of the levels that she encountered. Perhaps the last one was the most interesting one since the connection was through the daughter of a personal friend who when we first met had no idea that his daughter would be in the medical profession and therefore in a position to eventually help my mentee.

13. *Strengthening your portfolio* – Regardless of how you define "portfolio," it basically refers to your credentials put together in a composite form. It includes

your academic performance, your experience, your extracurricular activities, your letters of reference, your personal data, a personal statement, and actually anything else that defines you relative to a job, an educational program, etc. (essentially all of the items mentioned earlier). Certainly, you want to make your portfolio as strong as possible, especially when one considers the competition for positions. Thus, in addition to the items listed above, you can accomplish this in a number of ways:

- Make your presence known in the academic settings where you participate, e.g., class, seminars, both local and national, by participating, i.e., asking questions, responding to questions, answering questions, etc. However, do not monopolize the setting. For class, always present a positive image by attending all classes, being on time and paying attention, i.e., do not sleep, talk, and use electronic devices.
- Utilize the professor's office hours in a most positive way, and do it as a serious student, i.e., for questions about the lectures and/or course materials, not just to "brown nose." Nurturing the relationship with any professor is important, especially for mentors/letter writers (see below).
- Get involved in research with a faculty member, regardless of whether or not the research is exactly what you are interested in. The important aspect is to learn to think analytically, which research teaches you to do. Oft times an internship can serve this purpose.
- Identify mentors, i.e., people whom you develop a relationship with that usually goes beyond just the coursework and academic issues. A mentor has to really know you and what is going on in your life to truly help to guide you. As such, much of the onus on developing such a relationship depends on your efforts.
- Stay broad in your interests and correspondingly with your activities. This can be through specific student organizations related to your career choice, it can be an interest of yours such as sports, music, political issues, etc. or it can be at a community level, often through volunteering. This type of involvement reflects on who you are and is therefore most pertinent to what you want to do. You never know what area of interest may develop. My education and training is an excellent example of this. I received my PhD in endocrinology, working primarily in reproduction. As my career developed, I became very involved with, and committed to, minority issues. As it turns out, many of the minority health disparities have an endocrine basis. When receiving my PhD, I could not have anticipated that this area of research would relate so appropriately to the other area that I chose to pursue professionally.
- Prepare a resume (eventually to be replaced by a curriculum vitae) so as to provide a possible employer of your educational training, work experience, skills and abilities as well as your involvement with other activities. This is definitely your "marketing tool" and, as such, it should be concise, specific, and professionally well presented, i.e., no slang, abbreviations, small font, or grammatical/spelling errors.

- As you develop professionally, you will need to prepare a curriculum vitae, which highlights your educational/professional experience and achievements such as publications, educational positions/appointments, honors and awards, and references. It too must obviously be professionally presented and must be current. How one present themselves, both in written form and verbally, is critical in the creation of one's portfolio and all future networking.
- Always present yourself in the most positive professional manner, i.e., have a business card, use a professional email address, be aware of how you are portrayed on your Website, MySpace, Facebook, and even what message one gets when they reach your voice mail, e.g., do not have a prolonged period of music before one can leave a message.

In summary, there are some specific issues that adversely, and often preferentially, affect under represented and under served students. These include their backgrounds that do not provide the type of preparation for the next step toward their goal, racism at all levels with academia and society, lack of financial backing, lack of information regarding careers and their preparation, issues of self confidence and self esteem, and lack of mentors and role models. As mentioned earlier, there is no panacea, i.e., no one size fits all remedies; however, many of these have been recognized as impediments and obstacles and, as such, need to be addressed as the disadvantaged student proceeds to the next step. For example, coming from a deprived background, whether it is due to the community or the family structure, needs to be addressed as it is, i.e., something that the student overcame with dedication, perseverance, and anything else since that most definitely speaks directly to their personal strengths. Just to indicate simply that their background included this for the sake of including it will not work to one's favor. Similarly, overt acts of racism can be used to demonstrate one's ability to deal with, and get past, that experience. Other excellent resources for advice and tips in these areas include Adams 2002, Get up with something on your mind; Jones 2006, Just what the PhD ordered; Moore & Penn 2005, Finding Your North.

References

Adams HG (2002) Get up with something on your mind. Gemstones for Living. Marietta, GA.

Associated Press (2008) Some students take jagged path to graduation. Diverse Issues in Higher Education online. http://www.diverseeducation.com/artman/publish/2008/11/19

Elfman L (2009) AP results improve for other minority students, though Blacks still lag behind. Diverse Issues in Higher Education online. http://www.diverseeducation.com/artman/publish/2009/02/05

Jaschik S (2006) Momentum for going SAT-optional. Inside Higher Education online. http://www.insidehighered.com/news/2006/05/26

Jaschik S (2009) The impact of dropping the SAT. Inside Higher Education online. http://www.insidehighered.com/news/2009/03/26/sat

Jones SE (2006) Just What the PhD Ordered. Spare Not Publishing, Birmingham, AL

Limbach P (2003) Mentoring minority science students: can a white male really be an effective mentor?. Minority Scientists Network AAAS: Next Wave. http://nextwave.sciencemag.org/cgi/content

Mangan K (2009) Medical schools should re-examine admissions and training methods, experts say. The Chronicle of Higher Education. http://chronicle.com/daily/2009/01/10630n. Accessed 1 Jan 2009

Moltz D (2009) Promoting early college. Inside Higher Education online. http://www.insidehighered.com/news/2009/03/26

Moore EL, Penn ML (200[illegible]) Finding Your North. [illegible] LLC, [illegible]

Petchauer E (2008) But Professor? You're not white, you're German, right?. Diverse: The Academy Speaks – Diverse Issues in Higher Education online. http://diverseeducation.wordpress.com Accessed 8 Dec 2008

Chapter 5
Graduate vs. Professional School

There are obviously more than two possible career paths for science majors, although very often these are lumped into either graduate, i.e., Ph.D. or professional, e.g., MD, DO, PharmD, DVM, etc. This is done because these tracks are the more conventional routes that science majors tend to go. Also, it is somewhat natural since one route entails practicing a trade, e.g., medicine, pharmacy, etc., whereas the other represents a more classical academic route that usually focuses on teaching and research. Of course there are also the combined routes, e.g., MD/PhD, MD/MPH that some choose to take. Regardless, making such a choice involves a number of considerations and, as such, once again requires mentoring, specifically from individuals who are knowledgeable and experienced with the processes involved in these attaining these degrees and embarking into these careers, especially since the student may have very little if any direct experience and/or exposure to the actual requirements and workings involved.

Considerations for Making Career Choices

So now, going back to that initial decision, i.e., you are interested in the health sciences, do you want to conduct research as a scientist, e.g., research associate, professor, or deliver health care, e.g., physician, dentist, pharmacist? The first question that you must answer is whether or not you want to be involved in patient care since in the biomedical sciences, two major choices are doing research vs. having patient care. These are not mutually exclusive, e.g., nothing prevents an MD/DO from conducting research in addition to practicing medicine; in fact individuals with an MD/PhD routinely do research in addition to practicing medicine; however, individuals with a PhD cannot have direct patient care, except in specific instances such as a psychologist. Certainly, there are differences in the educational pathways taken for these degrees, even though there is significant overlap. In deciding to do research, one does need exposure and involvement in research at some stage of the education, since research truly represents a different experience than any other in the educational portfolio. In fact, the research experience is unparalleled in terms

T. Landefeld, *Mentoring and Diversity*, Mentoring in Academia and Industry 4,
DOI 10.1007/978-1-4419-0778-3_5, © Springer Science+Business Media, LLC 2009

of what it one learns and totally different than classroom learning. Because of that, the research experience can also lead to one deciding that is not the career that one wants. Regardless, that experience is most valuable for anyone interested in the health/biomedical sciences. (Many believe that a physician with research training is a better physician regardless of whether she/he ever does research again.) Additionally, the research experience may actually be in a clinical area and/or simply in a clinical setting. As such, the research experience provides a "look" at both career choices. Ultimately, one major difference is that as a health care provider, you will have responsibilities associated with patient care in one way or another. On the one hand, even as a pathologist, you deal with "patients" and their families. On the other hand, as a researcher in the biomedical sciences, you most likely will have an "association" with health care, e.g., if you are involved in research dealing with some aspects of disease which of course can be directly associated with health and disease. For example, you may be studying organisms involved with infections or, much more indirectly, you are examining basic science mechanisms that are integral to normal and abnormal physiological processes, despite the fact that your direct relationship with "the patient" is somewhat removed. This once again was the case in my career where in my research in the area of female reproductive endocrinology, I examined changes at the molecular level during the sexual cycle. Since the hormones involved were the same in the animal model as the human, my collaboration with a physician scientist who was also involved in the clinic allowed me to be involved indirectly with "patient care."

For those students who think that they want both, i.e., to be involved in research and also have patient care responsibilities, there are several options. One as mentioned is the combined MD/PhD. Once can obtain these degrees through combined programs around the country. In fact there are about 40 programs that are funded by NIH and defined as the Medical Scientist Training Program (MSTP). These are highly competitive programs, e.g., currently there are slightly less than 1,000 students nationwide that are in these and as such graduates of these programs are highly sought after. Many of these graduates choose to stay very active in research as Physician Scientists, some continuing to practice medicine and some not. Regardless, they can choose to be involved in patient care. Since they are funded by NIH, they actually cover the costs of both graduate and medical school, which based on the average debts of medical students, as discussed later, is a major consideration. For students not getting into an MSTP, for various reasons, but still desiring an MD/PhD, there are options. Many institutions will work very hard with students who desire both degrees so that a collaboration between the medical school and a department within the graduate school to make this happen. Graduates also are sought after; a major difference is that the student is usually responsible for the financial aspects of their medical education just as they would be if they pursued just the MD/DO. Finally, there is also an option for those individuals who want to do research and also have clinical responsibilities, but do not get the combined MD/PhD. They can pursue their medical degree, and along the way get research experience, e.g., between first and second years of medical school, hopefully coupled with some undergraduate research experience, and then even choose to do a fellowship upon graduation. Granted they do not have a PhD; however, with their clinical training combined with their research experience, there is nothing that prevents them from doing excellent research as a physician. Once again, this is most often a personal decision based on personal preferences as well as somtime simply the circumstances. An example was a student who was receiving his PhD at Michigan in Pharmacology and who was very interested in Third World health care. As he neared the end of his matriculation as a Pharmacology PhD candidate, he realized that he really could not accomplish what he had hoped in the area of health care with a PhD. As a result, he applied to medical school and received his MD degree. It might have taken him slightly longer by going through this route but, at the same time, he needed to see that was what he needed to do to achieve his long time goal. Moreover, being a PhD in pharmacology coupled with an MD provided him an excellent background for working in the area of Third World health care.

In addition, as mentioned in an earlier chapter with regard to researching one's options, there has been the development of a number of professional Master's degree programs in recent years. In fact, it is estimated that there are over 60 institutions now offering the so-called professional Master's degrees in a significant number of disciplines, e.g., bioinformatics, forensic sciences, geographic information systems (GIS) (Adams 2007, "Testing the waters"; Stripling 2008, "Scientist for a new age"). Part of this has been due to a recognition, particularly by industry that an individual could become a lab director with a Master's degree and further training "on the job" at the company, rather than the need to hire in with a PhD. Also, the science academy recognized that the road to a PhD and the subsequent tenured faculty position was long and tedious. As such, it was not a preferred choice for many who had taken a little longer to get their bachelor's degree and/or accumulated some debts associated with undergraduate loans. Often, underserved and under represented minority students fall into these categories. Along with these Master's programs is the development of bachelor's degrees in specialized major such as pharmacology. In both cases, i.e., the specialized bachelor's degree or the Master's do not prevent the individual from choosing to go on to an advanced degree in pharmacy, medicine, or a PhD. However, the marketability of the individual with this type of bachelor's and/or Masters degree is such that they may choose the workforce and have a career.

As in the cases discussed earlier, in making such a decision about your career, you really have to ask yourself, what really turns you on? That question is most critical in deciding on a career as it results in you having a career rather than a job. For example, if you state you want to help someone that does not eliminate doing research, you simply have to realize that your "help" will most likely be much more "long term" and certainly not as direct. With the Internet, it is possible to explore the possibilities extensively without ever leaving your home, library, or wherever you access the Internet. The initial query can be as simple as "careers in science." In fact, for a very comprehensive review of health careers, the Web site http://www.explorehealthcareers.org is excellent as it not only lists these options but also discusses the necessary educational preparation, the marketability, and even the salary considerations. Exploring sites such as this early in your decision making is most advisable. Then, you need to include exposure and experience in these areas as part of your education. For example, there is not just one medical degree and as such students need to know the difference. Most of society is familiar with the MD, i.e., medical doctor. Much of society and even academia is not so familiar with a DO degree, i.e., a Doctor of Osteopathic Medicine. What about an ND, i.e., a Doctor of Naturopathic Medicine. How about a Doctor of Podiatric Medicine? What are their qualifications? What do they do? What is the difference between an optometrist and an ophthalmologist? One has to know the answers to these questions if they are interested in that type of a career. Once they learn that, the next step is of course learning what is required to enter that career, what are the benefits, and does that fit with their goals and ideals. This Web site, http://www.explorehealthcareers/org, provides valuable information about all health careers, ranging from physicians to nurses to occupational therapists to biomedical researchers. It gives information not only about the type of preparation needed but also the marketing value and salaries of those in the fields and also includes listing of summer research programs making it one of the most comprehensive sites available.

Considerations About Conducting Research

Going back to getting involved in research, this can often be very straightforward, especially before such time that one chooses it as a career, i.e., going to graduate school. Basically, you identify a laboratory where they are conducting research and you offer your time and effort to be involved. Hopefully, this is a paid experience, depending on the financial status of the laboratory; however, if funds are not available and if you can afford to "donate/volunteer" your time, the experience will ultimately pay for itself. Also, quite honestly, the type of biomedical research in which you get involved is really not that important since: (1) The current overlap in methodologies, techniques, and therefore disciplines is such that the research experience itself is most valuable, and (2) the problem solving and analytical thinking that you gain will apply to whatever you decide to do in life. Some students actually get the opportunity to become involved in research while still in high school. For others it is when they are in college and fortunately even for those who are at a teaching institution with limited research facilities, there are plenty of opportunities during the summer to do research, almost anywhere in the country. As mentioned earlier, these are very valuable in one's academic experiences while planning for graduate/professional students. Not only do these programs pay you to do research, but quite often pay for you to get to and from the locations as well as room and board while you are there. Moreover, if the experience is a positive one, especially relative to the location, this site often represents a possible choice for graduate work.

As for "clinical experience," this is much more loosely defined, especially since students cannot be involved directly in patient care; however, being in a setting involved with patients and patient care can contribute significantly to career choices. For example, filing patient's records in a hospital or doctor's office is very different than shadowing a physician/dentist as they administer patient care. Similarly, being part of a research project that involves patients puts one much more in touch with the health care profession than a research project that does not. As an example, many students who say the reason that they want to become a pediatrician is that they "like kids." However, shadowing a pediatrician exposes the students to both the good, i.e., making the sick child well as well as the "bad," i.e., seeing children who are very sick and sometimes not able to be cured. Regardless, it is important for the students to get an overall feel of the practice of the profession before deciding on one.

Some serious misnomers and misconceptions still abound regarding researchers. A major one is that researchers are isolated and as such do not "deal with people." Granted, oft times, researchers do not interact with the public on a routine and regular basis as a doctor or dentists interacts with patients on a daily basis. However, scientists are more so than ever collaborating across laboratories and actually internationally due to the lack of defined boundaries previously associated with a discipline, departments, etc. Plus the problems facing the biomedical community require the combined efforts of scientists from all disciplines to solve them. Is there still a scientist who minimizes her/his interactions with others? Of course; however, that individual will enjoy very limited success in today's academic and scientific environment. There are even health care providers whose specialties minimize their "people interactions," e.g., pathologists, radiologists; however, still these individuals have "people interactions."

An important difference also between a researcher and a health care provider generally is that the health care provider practices a trade, actually much like an electrician. Now, that in no way should be taken as a criticism or in a degrading way. It simply means that the person practices a given trade throughout their career. For example, there are basically two ways to deliver a baby. Thus an Obstetrician does exactly that during their entire career. That in no way suggests that what they do is not most gratifying, exciting, and satisfying. That is just "what it is." A researcher, no matter how routine the technology/methodology gets is always asking new questions and exploring ways to address them. That is what research is all about. As one would expect, one's personality weighs heavily into making the decision as to which of these routes that one's eventually takes (and enjoys).

Another important point to be made in making such a career decision is having the knowledge as to what the opportunities are and how does one fit best within those. For example, many of the procedures and responsibilities that Nurse Practitioners and Physician Assistants have overlap considerably with those of a Physician. In fact, in many situations, the PA or Nurse Practitioner may be the primary health care provider, e.g., emergency room on the weekend when the physician is on call. Similarly, I learned of an Anesthesiology Assistant program, where the person works as part of a team with the Anesthesiologist and perhaps a

Nurse Anesthetist in the operating room. So, for the student interested in anesthesiology, it may be preferable to go to a 2-year Anesthesiologist Assistant program, thereby incurring less debt and being ready to "practice" sooner than later. At the same time, if the person wants to attain a medical degree and become an Anesthesiologist, then that is the path that they should pursue. The important point is that the educational pathways to get to these professions are very different in so many ways, including type of preparation, costs, length of time to degree, etc., so if a student knows this up front, the decisions are much easier and even more importantly, they are informed decisions.

Another very important point is that this decision, as important as it is, does not have to necessarily represent a final career decision. Individuals can always choose other careers, both inside and outside of the one that they choose initially. For example, for the first almost 25 years after receiving my PhD, I was involved primarily in biomedical research. Along the way, I became more involved in academic administration, e.g., Assistant Dean, Director of a Research Program, etc. As a result, I chose to pursue positions that were primarily administrative positions, still within the science discipline, but focused more on administrative programs rather than directed research. At the same time, I was able to combine both my research (grant) experience and my administrative experience to obtain research training grants for minority students. As such, I was able to change slightly the emphasis of my experiences while still staying within my originally chosen discipline. Others have chosen to step out of their discipline, e.g., going into law from science or perhaps business from science. In some cases, these moves have been totally outside whereas with others, they may be involved with either the legal or business side of science, e.g., patent law or the CEO of a technology company.

Financial Facts and Considerations

One component that must be at least considered in making these choices is that of finances as there is a big difference. The cost of obtaining a PhD in most of the sciences is nothing, at least academically since these programs pay a good stipend, e.g., $25–30 K along with tuition and benefits. Thus, on the one hand, many students graduate from these graduate programs with only debts incurred relative to their lifestyles, e.g., car/house payments, etc. On the other hand, students choosing a health profession, in most cases, are going to be in debt following their matriculation. For example, the average debt of a physician upon finishing medical school is somewhere is the vicinity of $150–170 K. Granted, eventually that debt can be paid as the salaries of physicians are quite good. However, the individuals do not reach those high salaries until they are well into their practices, e.g., residents earn about $40–50 K. With that being said, the key consideration is "what do you want to do; what turns you on?" If it is the practice of medicine or dentistry, then go for it knowing that the debt will be there and eventually paid. In other words, do not choose a PhD route solely because of fiscal considerations, as that reason alone most likely will not be enough to sustain you throughout your career. Similarly, do not choose not to go into a health profession solely due to the amount of debt that will be incurred, but rather accept that fact because you want to practice medicine.

Facts About Getting a PhD and What Then?

A number of recent studies have addressed the retention, or more appropriately the attrition, of graduate students in the sciences. On the basis of many of the concerns and issues facing minorities, it is not surprising that the attrition rates for women and minorities are higher than those for majorities, especially considering that the overall attrition rate for graduate students is about 50%. (This is considerably higher than the attrition rates for professional schools for a number of reasons, including the more "unstructured" graduate programs and the sheer numbers that enter graduate programs every year.) For women and minorities in science, an obvious reason for this is the numbers alone, i.e. fewer women and minorities are in the graduate sciences programs (and always have been).This of course results in an isolationism, and in some cases, an alienation as discussed in an earlier chapter. With that of course comes an unfavorable, even sometimes, an intolerant, and insensitive environment. With the many pressures associated with succeeding in graduate school, such as grades, research productivity, and general adaptability, an uncomfortable environment is sometimes the "straw that breaks the camel's back." As will be discussed in a later chapter, a contributing factor toward this environment is a similar lack of minority faculty members (and those nonminority faculties who are sensitive to the minority students and their issues). When these are taken together with other educational system issues such as often the types of K-12 preparation of these students, it is not surprising that the attrition rates are as high as they are. As for how to address and improve these, there are once again many possible solutions since as stated very early in this text, the places of the leaks (attrition) have been long identified as has the ways to fix them; it is the willingness to put the efforts, time, and monies into the "fixes." For example, until such time that there are indeed sufficient numbers of mentors who can effectively mentor minority students, there has to be a commitment from the very top of the organization to provide whatever is necessary to do this. An excellent example of this is Dr. Freeman Hrabowski and his commitment as President of UMBC, as described in a recent paper entitled "The University as Mentor: Lessons learned from UMBC Inclusiveness Initiatives" (Bass et al. 2008, CGS Paper Series on Inclusiveness). Similarly, Spelman College has an outstanding record of sending Black women graduates onto PhD programs in the sciences (White 2008, 11/16/08 *Atlanta Constitution* "Spelman Students sparkle in sciences"; Thomas 2008, "Spel-bounding: All female Spelman college ranks no.2 sending Black graduates onto PhDs in Science and math"). In both of these cases, the commitment by the institution is a critical component in their success. Personally I can attest to the type of commitment from my experience of giving a talk on Career Choices for Minority Students in the Sciences both at Spelman College in 9/08 and Morehouse College in 2009. Both talks were at 5 PM on a Friday afternoon and the room was not only packed with students but also many stayed after the talk, again late on a Friday afternoon, to obtain further information. In contrast, many campuses are virtually deserted on Fridays let alone at 5 PM. Also, importantly, the groups were Black women (Spelman College) and Black men (Morehouse College) in science majors, which are both groups which deserve attention as under represented in STEM. Interestingly, despite these turnouts, I will talk later about how women may be affected sometimes disproportionately relative to staying in the "business of academic science" as well as mentioned earlier, the dearth of minority males in education but particularly the sciences.

Still another aspect that is being considered heavily relative to making the choice to become a graduate student is that of having a "balanced life." In fact, in a recent survey with more than 8,000 graduate students across the University of California system (Mason et al. 2009a, b, "Why Graduate Students Reject the Fast Track"), many highly qualified students are choosing not to enter PhD programs, primarily due to the reputation that academics do not have well balanced lives, i.e., little room for a satisfying family life. Much of this is not only due to the fact that times are different but also since more women and minorities are considering such options. And, although this is very much the case for positions in research-intense universities, it is not a problem only for these types of institutions. When we combine these concerns with those that many have regarding science and factor in the issues for minority students, fixing the pipeline and successfully addressing the under representation is an overwhelming and time-consuming task.

Finally, a consideration for choosing to enter a PhD program at least in current times, i.e., 2009, is the fact that there is a "PhD Admissions Shortage," according to an article by Jaschik (2009) in *Inside Higher Ed* (3/30/09). This is not surprising considering the economic status in 2009, and especially since many PhD programs, and almost all in the biomedical sciences, pay a stipend as well as tuition and other costs. Admitting fewer students, regardless of application quality, is a necessity at a number of institutions. Institutions that were mentioned in the article included Emory University in Atlanta, the University of South Carolina, Columbia University and NYU. The cuts in admissions ranged from as little as 10% to as much as 40%, with some across the board and others in selected departments. An obvious impact could certainly be seen down the road with regards to less teaching assistants and new professors as well as less research, which is the "heart and soul" of science PhD programs. At the same time, some argue that the current job market is not as strong as one would like for new PhD's, especially in academia and, as mentioned in other sections of this book, even more so for minorities. This is an issue that has to be observed very closely and is obviously tied very tightly to the country's economic status and the prioritization of education.

It is also important for students who choose to obtain a PhD in the sciences to realize that once they attain the degree, they will usually have to seek additional training, often referred to as a "postdoc." This is of course short for postdoctoral and is further training, usually focused on research that provides the recent PhD with a broader spectrum of research techniques and methods. As such this training can be in a very similar discipline as the one where they got their PhD, which has the advantage of them possibly being able to accrue more publications or it can be in a very different area, which of course greatly broadens their research expertise and in many cases makes them more marketable for a larger array of positions. The length of time for these postdoctoral positions varies from a year, which is fairly uncommon, to as much as 4–5 years. Many are in between these times, e.g., 2–3 years. Another possible advantage of a postdoctoral position is obtaining some teaching experience, since many PhD programs particularly in the biomedical sciences do not often provide the candidate with much teaching experience. This can prove most valuable for students who want to continue doing research but want to balance that with some teaching and are particularly relevant to those who want to teach at a smaller institution, e.g., an MSI where teaching is the main focus and facilities and teaching support is limited. To address this, NIH (NIGMS) actually developed a program Institutional research and academic career development (IRACDA) in which the participants, as postdoctoral fellows, did research at a research institution while also teaching at an MSI. Regardless of which "route" one takes, the individual should definitely plan of spending some time in a postdoctoral position not only for the reasons mentioned already relative to their training and credentials but also because now most candidates for positions have postdoctoral experience and thus without it a candidate is less likely to be considered for a position.

From the PhD and subsequent postdoctoral experience, many will then compete for a faculty position. Here again, there are certain things that one must be aware of. First, to eventually obtain security with the position, the candidate should see a tenure-track position, knowing that obtaining tenure usually involves a period of 5–6 years. Since tenured positions do represent a long-time commitment by an institution, many institutions have other tracks that are nontenure track where the appointment is on a temporary basis, often tied to the individual's ability to obtain and maintain funding (or being supported on someone else's grant monies). In fact, some institutions have chosen to no longer offer tenure. In either case, the individual may be appointed as a part time faculty member, where the support if actually from the institution, e.g., state monies for a state school. Although the monies are potentially more "secure," they are dependent upon the state's economy, and as a result many part time faculty commitments are 1 year at the most. This means that one should pursue tenure-track positions, if one wants the security associated with one position, and because of this, the competition for these positions is significant. The result is that the candidate's portfolio must be especially strong in all areas, especially research, to be competitive, again a reason for postdoctoral experience.

Thus, the best advice for all students who are interested in a career in the biomedical field is to try to get research experience. Even if one does not conduct research later on in their career, the research experience would have taught them some valuable lessons about dealing with not only issues associated with science but also with society as well. In other words, thinking critically and analytically will help anyone cope better with problems that they face in the sciences and in life. Along those same lines, getting exposure to clinical settings can really result in "eye-opening" experiences that can either turn a student on further to what they believe that they want to do or possibly totally turn them off to it. Unfortunately, at many campuses, particularly small undergraduate and often minority, ones, the opportunities to do research or get clinical exposure are limited, especially the former as research requires facilities, funding, etc., as opposed to clinics and hospitals, which are much more available. If this is the case, the student has to explore opportunities outside of then "comfort zone" through for example summer research programs. Actually these types of programs are advantageous in other ways in that they expose the students to a brand new environment; in fact one that if the situation is "right" may lead to them applying there for their postbaccalaureate education. As an excellent example of this, I strongly encouraged a Hispanic male who was in my research training program to apply for a summer program outside of our institution. He was accepted at an excellent Big Ten university in chemistry. As part of his summer experience, he visited another of the Big Ten campuses for the summer end conference. When the time came for him to apply to graduate school, he applied to both of the Big Ten schools and in fact ended up enrolling and eventually getting his PhD in chemistry at the school which he visited that summer as part of the weekend conference. I am totally convinced that he would not have applied to either of those Big Ten schools if he had not spent the summer at the one.

The bottom line is to be informed, whether it comes through advisors, mentors, the Internet, friends, family, or whatever source, as this is the most important consideration for making such a choice. Unfortunately, this does not eliminate the possibility of making the "wrong" decision, but it does eliminate making the wrong decision based on not knowing what you needed to know. That is the important issue. Do your homework and ask questions – that is how you become informed.

References

Adams JU (2007) Careers and grad programs for B.S/M.S. Scientists: Testing the waters. Science http://sciencecareers.sciencemag.org/career_development/2007/09/28

Bass et al (2008) The University as Mentor: Lessons learned from UMBC Inclusiveness Initiatives. Council of Graduate Schools: Volume one in the CGS occasional paper Series on Inclusiveness

Jaschik S (2009) PhD admissions shortage. Inside Higher Ed http://www.insidehighered.com/news/2009/03/30

Mason MA, Goulden M, Frasch K (2009a) Why graduate students reject the fast track. Academe Online (AAUP) http://www.aaup.org/AAUP/pusres/academe/2009 Accessed 1/20/09

Mason MA (2009b) Balancing Act; A bad reputation. The Chronicle of Higher Education http://chronicle.com/jobs/news/2009/01/2009012701c Accessed 1/17/09

Stripling J (2008) Scientists for a new age. Inside Higher Ed http://www.insidehighered.com/news/2008/07/14

Thomas CR (2008) Spel-bounding: All female Spelman college ranks no.2 in sending Black graduates onto PhDs in science and math. Diverse Issues in Higher Education online http://www.diverseeducation.com/artman/publish/2008/12/08

White G (2008) Spelman students sparkle in sciences. Atlanta Constitution 11/16/08 www.explorehealthcareers.org

Chapter 6
The Application Process

The culmination of the experiences that the student gains while they are progressing through the system will be represented in their application to the professional or graduate school. As such, this step is a crucial one and requires much thought, planning, and effort. There is no shortage of schools to apply to, so efforts must be expended as to where to apply and why, how many is the right number, and, just like the importance of individualization with regards to mentoring, each individual school should offer something that "works" for each individual, whether it be something as straightforward as the weather/climate to the school's reputation (as long as it is based on "legitimate history" such as students from the person's school being successful there). Moreover, it could be the type of curriculum that is offered, which is especially relevant to medical education with the change in recent years or the research being conducted or the type of programs offered, e.g., specific PhD discipline vs. the "umbrella biomedical sciences" degree. The point is that the student must make a sincere effort to identify those parameters that are most important to them as an individual in choosing the schools to apply to and being able to indicate that as part of the application including the interview process. Input from experienced mentors is crucial at this stage and for this reason.

Application Facts

Once the decision is made as to which educational pathway is chosen, based of course on the chosen career, then comes the preparation of the application and I say application in the singular since no matter how many programs/schools that you apply to, the information provided will be very similar. For this reason, it is imperative that your application be flawless, i.e., there are not only no mistakes but also it states everything that they have requested, and everything that you can provide to make you the best applicant possible. Quite honestly, due to the number of applications for the number of positions, admissions committees will usually look for negatives, i.e., red flags that "jump out." Once these are observed that particular

T. Landefeld, *Mentoring and Diversity*, Mentoring in Academia and Industry 4,
DOI 10.1007/978-1-4419-0778-3_6, © Springer Science+Business Media, LLC 2009

application can go from the initial pile into a "maybe" or "lower priority pile," and this is the last thing that you want to happen to your application, as sometimes committees get back to that group of applications and sometimes they do not. Granted there are some things that you can do nothing about in your application, e.g., GPA, standardized test scores; however, there are many other components of that application that you do have control over and those are the parts that you make sure to provide the strongest picture of you and your credentials, especially as already mentioned, the personal statement, which is where you can actually address any of those possible "red flags." Two of these are of course the personal statement and the letters of recommendation. The statement is totally under your control as you are the one writing it. It should not be "done" until you are totally satisfied that it says what you want (and you have had others see and comment on it). Although you cannot directly control the letters of recommendation, as discussed earlier if you successfully nurture the relationships with those mentors who will be writing the LORs, including seeing the copies of them, you will "have control" of this component of the application as well. With that, the application can be as strong as possible before submission.

The choice as to where to apply involves information gathering just as with the career choices and of course the sources for this information are the same, i.e., advisors, mentors, Internet, institutional Web pages, friends, etc. For example, advisors/mentors often have first-hand knowledge of schools where they have had, in one way or another, interactions with in the past. Web sites for organizations such as the http://www.AAMC.org and the http://www.AACOM.org, just to list two for medicine, have an enormous amount of information about the institutions that offer the professional degree as well as including standardized test scores and GPAs of applicants and matriculants, minority affairs activities, curricular information, enrollments, and much more. As such, the student needs to identify how she/he fits within these programs considering all of these factors but highlighting certain ones that are most important to them. Even the consideration of the climate where the school is located becomes very important in the decision-making process. Very often it is helpful to list these and then "score" them for different institutions with the total score hopefully representing those that are most favorable and as such most desirable. As an example, there are a number of medical schools in the Caribbean; however, only a few of these currently have accreditation from the United States, meaning of course that when a student finishes in one of those schools, they can meet the requirements of the US hospitals for residencies, which is critical to their career. Certainly, although the "relocation" is temporary, i.e., usually 2–5 years, that is a sufficient amount of time that requires as much optimization of "conditions" beforehand as possible, even for example living on an island in the Caribbean many miles from their home.

It was stated that much of the information requested by all applications is the same and also as stated this is an advantage as it permits the applicant to concentrate of essentially one application. In fact, many of the professional schools have a centralized application process, which means that the student completes one application, identifying the schools where the application is to be sent. The costs include an initial cost with additional costs due to the number of schools where the application is sent. Most of the information will come from the student's portfolio, which they have hopefully maintained through their education pursuits. Even if they have not, the information is either readily available or easily gathered and collated. This information includes GPAs, various activities, letters of recommendation, and personal statement. Very often, schools will request information specific to their programs. However, again, this information is often similar across institutions.

Interviewing Tips

The interview is obviously the key component in the application process, as all of your application materials are designed with the goal of securing an interview. It is then your actions at the interview that are most likely to determine whether or not you receive an acceptance, i.e., the best candidate does not always get the acceptance but rather many times it is the person who is best prepared for the interview. Tips for best preparing for the interview as well as performing well in the interview include the following:

1. If possible, participate in interviewing workshops and have a "mock interview." Both of these will prepare you for the types of questions that you may be asked as well as the types of answers that are best. Since body language and movements are important, it is even better if the mock interview can be video taped. Also, there are copies of documents e.g., "the 100 most asked questions in interviews" that can be helpful but these are not nearly as valuable as a mock interview.
2. Prepare yourself regarding information about the institution/program/organization, as the more that you know the more that you impress the interviewer as to your interest. This information is available on Web sites, catalogs, and other published documents.
3. Be sure to express your strengths as highlighted in your personal statement as the interviewer really wants to know "who you are and what you can contribute as an individual to the program." Be confident but not arrogant. Also enthusiasm is a most positive feature.
4. Pay attention to the interviewer so as to best understand his/her "agenda/motives" and answer the questions as completely as possible.
5. Do not talk "too much" and do not ramble. Keep your answers succinct but as mentioned before, as complete as possible.
6. Do not answer every question as quickly as possible, especially if you are unsure of your answer. You can ask the interviewer to repeat the question or even state "That is an interesting question, will you give me a minute to think about it?." Because once an answer is given, it is virtually impossible to take it back.
7. Refrain from negativity, especially relative to professors, institution, jobs, etc. A positive attitude and positive responses are always viewed "positively." (Not surprisingly!)

8. The interviewer has significant input on the decision for acceptance; do not underestimate that fact.
9. Be prepared to discuss almost anything. For example, when I interviewed a student for the combined MD/PhD program, we spent much of the time discussing the book Their Eyes Were Watching God as it was a book that both of us enjoyed and admired. On the surface that may not have seemed to have direct relevance to an interview for a slot in a sciences graduate program, but it certainly spoke directly to the interests and character of the interviewee.
10. Always maintain eye contact with the interviewer. Not doing this suggests a lack of confidence, untruthfulness, or both.
11. Have questions ready to ask, especially when asked. Otherwise, a lack of interest and/or self esteem may be perceived.
12. By using your knowledge about the institution, refer to positive aspects/strengths and how you can fit into these. A question that you should have an answer to is: "Why did you apply here?"
13. One question that is appropriate is: "What is the next step in the process relative to your application?"
14. It should be a given that you dress appropriately, you permit no distractions such as a cell phone and that you arrive early.
15. End the interview by acknowledging the interviewer's time and efforts.
16. Sending a thank you note to the interviewer at a later date is not required but very often is a "nice touch."

Do not underestimate the importance of the interview and prepare accordingly.

References

www.aamc.org
www.aacom.org

Chapter 7
Faculty Advising and Mentoring

So who becomes a mentor and why? Once a mentor, who does one mentor, e.g., does a biologist mentor a chemistry student? Or as is particularly relevant to the topic of this book, does a nonminority mentor minority students? And, if that is the case, how does one do it effectively and successfully? The answer to all of these questions is that the key to good mentoring (and yes there is bad mentoring!) is individualization. No two students are the same regardless of whether or not they have the same major, are the same gender, the same ethnicity, or even from the same family. Recognizing the differences and addressing them is what a good mentor does. And, although there is training for mentors, many mentors learn from their own experiences and that is why we tend to think of mentoring as a process by an "older" individual for a "younger one," which is not always the case. Moreover, to address the minority/nonminority issue, it is critical for nonminority mentors to be sensitive to, familiar with and understanding of the issues that may not be in their experiences but are paramount for minorities.

Mentoring vs. Advising

First, students need to recognize that advising and mentoring are not the same and as such not all advisors are mentors and vice versa. This is important to note since advising is something that almost all faculty do; in fact, at most institutions, it is a defined part of their job responsibilities. And, as important as it is, many times, especially with certain students, advising can often be basically a "rubber stamp," i.e., there are the courses that one needs to take, and do so in a certain order as dictated by the necessary prerequisites. As such, an advisor will approve and disapprove, and if the latter, make recommendations. As stated, this is important but is very different than mentoring, which as important as it is, is not usually a part of the job assignment (more later) and requires much more time and effort. As a result, many faculty members do not have, or take, the time to do mentoring. Moreover, when one considers the importance of this especially for students that are nontraditional, e.g., minority, there is a real need for students to seek out effective and caring mentors and,

T. Landefeld, *Mentoring and Diversity,* Mentoring in Academia and Industry 4,
DOI 10.1007/978-1-4419-0778-3_7, © Springer Science+Business Media, LLC 2009

in turn, for faculty to decide if they indeed want to make this commitment. As mentioned earlier, mentoring has been demonstrated to be the most effective tool in the successful development of a minority student. However, due to the long-term problem of the underrepresentation of minorities, a severe lack of minority mentors obviously adds significantly to the lack of effective mentoring of minorities, at all levels, which in turn, contributes to the underrepresentation! This is further complicated by the fact that many nonminority mentors have chosen not to sensitize or familiarize themselves to such issues and, as such, cannot effectively mentor the students in those areas where assistance in mentoring is sorely needed. In other words, traditionally, individuals will mentor in those areas where they feel most comfortable and/or they have the most experience/expertise (Evans, J. 2008 Mentoring magic: How to be an effective mentor, tips from two highly successful principal investigators). This is natural and usually works. However due to the paucity of individuals with experiences and expertise in the area of minority affairs, there are "not enough to go around" and, as a result, many minority students are not mentored, often resulting in them dropping out or in some way being lost in the system. For example, in most educational programs, the attrition rate for minorities and women are higher than that for others. As an indicator of how important mentoring is, at the graduate student stage, one only has to look at one of the recommendations of the Carnegie Initiative on the Doctorate study that included, 84 PhD-granting departments in six different fields (Wasley 2007, "Carnegie Foundation creates new 'owner's manual' for doctoral programs"). The study recommended that "doctoral programs adopt new structures that allow students to have several intellectual mentors and come to think of mentorship as less an accident of interpersonal history and as more a set of techniques that can be learned, assessed, and rewarded." From my experience, the same could be applied to all levels of the educational pipeline.

Underrepresented Minority Mentors

One "fix" of this problem is, of course, increasing the number of minority scientists so that there are more potential mentors. As mentioned in an earlier chapter, the process for achieving this has been extremely slow and despite more programs designed to increase diversity, attaining sufficient numbers of minorities to in turn mentor minority students resulting in more minority representation in graduate/professional programs and ultimately biomedical and health careers is quite "far off." Still, these efforts must be made. Until the numbers of minority scientists and therefore minority mentors and role models increase significantly, there are ways to accomplish at least the exposure minority students to existing minority scientists. For example, I was involved in the development of a course on my campus designed to expose our students, and in particular, minorities not only to research being done by minority scientists but exposure to many of the trials and tribulations that those scientists endured to get where they were. In other words, part of the speaker's talk and discussions obviously included points about dealing with the system as a minority scientist. In addition, of course, the students now saw "people that looked like them" who were indeed accomplished and had also experienced many of the same things that they were experiencing. Another added benefit of this seminar series is that the invited speakers also met with faculty and administrators, not only exposing them to same the "real facts" but also doing it from a perspective of an accomplished scientist, coming through the system from often a disadvantaged standpoint. Very often, as the course director, I would actually focus on inviting young developing minority scientists so that they could really relate well to the students while at the same time gaining valuable experience in presenting their work and adding to their CV. Most often, of course, they were invited and advertised primarily due to their research since focusing on the other component would not necessarily strengthen their vitae in the eyes of their institution. Still another way to accomplish this, i.e., expose the students to role models and mentors is to have them attend national conferences, and in particular, ones that focus on minorities in science. Two excellent ones held every year that highlight minority students involved in research are the Society for the Advancement of Chicanos and Native Americans in Science (SACNAS) and the Annual Biomedical Research Conference for Minority Students (ABRCMS). Not only are the students exposed to outstanding scientists, both minority and nonminority, but also almost 2,000 other minority students who look like them and are committed to advancing in the sciences. This experience is invaluable for so many reasons, including of course the networking that is accomplished there. It also gives them confidence for when they then attend a national scientific society meeting and are truly "in the minority."

Of course, not all minority scientists will be, or should be, expected to be mentors for other minority scientists, in part due to the fact that not everyone can mentor well. Plus, the demands upon the minority scientists are usually major with mentoring just being one of those responsibilities. At the same time, what about those nonminority scientists that are in positions to mentor young scientists? Is there any reason that they cannot be effective mentors? The answer is of course, no, there is not a reason that they cannot and, in fact, there have been a number of excellent examples of such individuals. However, these numbers are also small and for very practical reasons, i.e., one mentors most comfortably and usually most effectively in areas where they have the most in common and are most "at home," whether it be due to their background, their discipline, their gender, or their ethnicity. And, since there are plenty of white scientists, issues relating to ethnicity often is the area where mentoring is most needed. This of course means that nonminority mentors must mentor minority students to meet the need. Is this a problem? Again, referring to the numbers alone, coupled with the level of comfort, it has been and most likely will continue to be.

Sensitivity and Cultural Considerations

So how does one becomes more knowledgeable and, as such, more sensitive to an area where they have little, if any, experience, exposure, and/or possibly even interest? Well, considering the changing demographics and the commitment to diversity, there are a number of sensitivity training sessions provided almost everywhere. In fact, very often, participation in these is financially supported by the organization/institution as the results are designed to make for a better environment for all. These can work, since in some cases, exposure to the different cultures and the nuances associated with them is enough to stimulate a change in the person and in their resultant mentoring. Of course, most importantly, the participant has to decide to apply what they have learned in their workplace. Even after deciding this, it is still often difficult as not all possible situations/scenarios would have been covered in the training sessions and being faced with a "real-life" situation is very different than the staged ones in the training sessions. However, the important component is the desire to face these situations with a different approach, one where the mentor takes themselves out of their familiar setting and transfers themselves into another. This is never easy and even over time there will still continue to be situations where the mentor has difficulties. However, making the commitment and the effort to do this is the first important step in effective mentoring of nontraditional students (Limbach 2003; Mentoring minority science students: can a White male be an effective mentor; Petchauer 2008; But Professor? you're not white, you're German, right?). And actually, in some ways it is really not that much different from many mentoring situations where the mentor and the mentee are "different" in a number of ways, e.g., a male mentoring a female, a biologist mentoring a chemist, etc. In most all of these cases, the mentor has to look at the mentee in a special way to effectively provide mentoring and does so usually by "projecting," to some degree, themselves into another role, most often one which is unfamiliar to them. In fact, the key to effective mentoring is treating all mentees as individuals and as such recognizing their unique qualities, strengths, and flaws, i.e., personalization, as opposed to those who often pride themselves on saying that they are "color blind." How can someone deal with individuals as individuals if they do not see them as they are, whether it be their gender, ethnicity, skin color, etc.? In fact Nealy's (2009) article in *Diverse Issues in Higher Education online* (3/20/09) contains the following quote from Paul Kivel's book *Uprooting racism: How white people can work for social justice:* "To avoid being called racist, [White Americans] may claim that they don't notice color or don't treat people differently based on color. [But] it is not useful or honest for any of us to claim that we do not see color." This was further reinforced in a survey published in the online version of the journal *Psychological Science* where researchers found that Whites who subscribe to an ideology of colorblindness in the workplace cause their minority colleagues to feel less committed to their work. On the other hand, when Whites champion multiculturalism, their minority peers feel more connected to their jobs. Dr. Victoria Plaut, the study's chief author, states that "Minority employees sense more bias in these allegedly colorblind settings." As a result, she recommends that more emphasis should be placed on creating environments that recognize and celebrate racial and ethnic differences, not going the other way, i.e., "colorblindness." Granted in mentoring situations there may be overlapping themes, issues, concerns, but how each individual deals with these depends on them as an individual. The mentor must recognize and appreciate this in mentoring the person. When it comes to the issue of mentoring minorities and therefore recognizing and acknowledging ethnic differences, most assuredly this is true, probably even more so than for any other example. This is due in large part to the country's preoccupation with "race" which incidentally is not by any means a scientific term but rather a sociological one. As such, to accomplish such a "transition," it is best stated by Woodson (1990) in *The Mis-education of the Negro* when he stated "It is all right to have a white man as the head of a Negro college or to have a red man at the head of a yellow one, if in each case the incumbent has taken out his naturalization papers and has identified himself as one of the group which he is trying to serve. It seems that the white educators of this day are unwilling to do this and for that reason they can never contribute to the actual development of the Negro from within. You cannot serve people by giving them orders as to what to do. The real servant of the people must live among them, think with them, feel for them, and die for them." This is most definitely true for effective mentoring of minority students and therefore represents a problem due to the shortage of individuals committed to doing this. As such, this has been and will continue to be a problem due to the practice of white privilege, something that many are unwilling to acknowledge, let alone "give up." There are certain examples where nonminority faculty members have most effectively and successfully mentored minority students as well as nonminorities who have "given up" some of their white privilege to make a difference in the world of discrimination and biases (Wise 2005, "White like me: Reflections on race from a privileged son"; Wise 2009, "Between Barack and a hard place: Racism and white denial in the age of Obama"; Thompson et al. 2004, "White men challenging racism"). Personally, I am one that has done both, i.e., mentored successfully a large number of minority students as well as given up some of the white privilege advantages along the way. In the book, *Confronting Authority: Reflections of an Ardent Protester*, Bell (1994) speaks of the consequences that one has to deal with when "fighting the system," which is what sacrificing white privilege is all about. Unfortunately, the number of these individuals remains relatively few and, as such, the effort of mentoring minorities by nonminorities continues to be a real problem.

Thus, besides participating in sensitivity sessions, a person has to individually make an effort to become sensitive and aware of issues of others. As mentioned, this has to involve immersing themselves into another culture as much as they feel comfortable. This does not have to mean totally relocating or even changing completely one's social habits; however it does mean becoming more a part of another culture than one previously experienced. This effort serves many purposes, one of which is that of a commitment to learning about others so that one can better understand and mentor. In fact, one only has to look at the list of reasons, cited earlier (Hayes 2009, Maryland AG offers legal guidelines for increasing diversity in state's universities), as to why diversity is important. This alone is critical since often just by the student recognizing this type of commitment, a different type of mentoring relationship can develop, beginning with a respect that may not have been there initially. Then, with continued "immersion" and repeated mentoring, this comfort level, by both the student and the mentor, becomes almost commonplace.

Why is this so important, i.e., that we identify more mentors from minority groups and/or are sensitive to minority issues? It is well established that students look at mentors as role models regardless of their relationship to them relative to gender, ethnicity, discipline, etc. However, at the same time, when the mentor "looks like them" or at least can relate to them relative to issues that are intrinsic to the student and not always so for the mentor, the mentee is much more comfortable and receptive. Both of these contribute to a positive environment, which is critical for effective mentoring.

References

Bell D (1994) Confronting Authority: Reflections of an Ardent Protester. Beacon Press, Boston, MA

Evans J (2008) Mentoring magic: How to be an effective mentor; tips from two highly successful principal investigators. The Scientist 22(12):70

Hayes D (2009) Maryland AG offers legal guidelines for increasing diversity in state's universities. Diverse Issues in Higher Education online http://www.diverseeducation.com/artman/publish/2009/03/16

Limbach P (2003) Mentoring minority science students: Can a White Male really be an effective mentor? Minority Scientists Network AAAS; Next Wave http://nextwave.sciencemag.org/cgi/content/2002

Nealy MJ (2009) New study: Colorblindness has negative effect on employees. Diverse Issues in Higher Education online http://diverseeducation.com/artman/publish/2009/03/20

Petchauer E (2008) But Professor? You're not white, you're German, right? Diverse issues: The Academy Speaks http://diverseeducation.wordpress.com Accessed 8/12/08

Thompson C, Schaefer E, Brod H (2004) White men challenging racism. Duke University Press, Durham

Wasley P (2007) Carnegie Foundation creates new owner's manual for doctoral programs. The Chronicle of Higher Education http://chronicle.com/daily/2007/12/868n Accessed 12/4/07

Wise T (2005) White like me: Reflections on race from a privileged son. Soft Skull Press, Brooklyn, NY

Wise T (2009) Between Barack and a hard place: Racism and white denial in the age of Obama, City Lights Books

Woodson CG (1990) The Mis-education of the Negro. Winston–Derek Publishers, Inc, Nashville TN

Chapter 8
Surviving in Professional Careers

Once an individual is into career, whether it be as a professor, a practicing physician, or a pharmacist, it may seem that much of the struggle is over and, relatively speaking, it is. However, this does not mean that there are still not obstacles that one must face, especially for women and minority scientists. For example, although the number of minority women entering science careers has increased in recent years, the attrition rates for minority women are higher than for men. This, of course, is due to a number of factors, some of which are similar for minority men as well. For example, the low numbers of minorities in the ranks make "the fit" difficult regardless of gender; however, since men still dominate the faculty ranks, women, and in particular, minority women often find the environment quite uncomfortable, sometimes even hostile. Add to this the fact that women still earn less despite having comparable credentials, it is not surprising that retention is an issue. Paramount to the issue of retention is the need to demonstrate success in one's discipline whether it is through teaching excellence, research productivity, or extensive service to the school, community, and the discipline. In fact, for many, all three of these are essential in demonstrating success as well as maintaining it.

Things That Must Be Done to Survive

So now you have made it! You have a position at a college or university. Or you are at a pharmaceutical company. Or you are a practicing dentist. No problem. Right? Well, not exactly! Definitely you have moved to the level to which you have aspired for some time, that is, through all of the examinations, the term papers, the research projects, etc. However, you must still be very much aware of what you need to do to stay on top of your "game" to not only stay in your chosen profession but also be successful in it. You will find that there are many things that you still must learn – things that you were not taught in the classroom or the laboratory. In fact, you probably have only scratched the surface in learning what it takes to "play the game and play it right." At the same time, some of the very same things that you did in preparing for your career apply once you enter into it. For example:

T. Landefeld, *Mentoring and Diversity*, Mentoring in Academia and Industry 4,
DOI 10.1007/978-1-4419-0778-3_8,

1. You must continue to add to your portfolio, which not surprisingly includes a constant updating of your curriculum vitae, taking courses or training to keep you "up-to-date," staying visible in the right circles, participating in professional organizations and societies that are most relevant to your chosen discipline as well as occasionally reviewing your personal statement, which, of course, still speaks to your goals and objectives. What if those goals and objectives change? This is not a problem and quite frankly they probably will either slightly or possibly significantly change, based, of course, on your path once you are in your profession. Regardless of which of these scenarios is the case, you will use your background preparation plus your experiences to make this decision.
2. Directly related to the previous point is the fact that one has to prioritize their interests/efforts both in their current position as well as their future aspirations. For example, all faculty members have a responsibility to serve on committees, at all levels throughout the institution and although some faculty see this as only a necessary evil, others embrace it, especially when it is obvious that larger effects can be seen when one works at a programmatic level. In fact, in some cases, these faculty members will choose to continue to pursue various administrative positions so as to be able to more effectively make a change. This is not for everyone but as one continues to find their "niche" and actually talk about it in their personal statement, their portfolio becomes one that is attractive for that type of a position. This was actually the case with me as I eventually entered into full-time administration so as to make a difference programmatically.
3. You will continue to network, that is, create those "hook-ups" that can be used to progress your career as well as to possibly move into another direction. There is always going to be someone who can assist you in whatever you choose to do, whether it is by advising/consulting or introducing you to the "right person or persons." In the sciences, collaborations are an essential part of almost all successes and these collaborations are usually the most direct route to "hook-ups."
4. Directly relevant to point #2, you must continue to keep those individuals who have supported you in the past with letters of recommendation apprised of your current situation and progress as well as "nurturing the relationship" with new individuals in your current setting. Nothing is worse than a letter of recommendation written by someone who knew you "back then" and has no new information on you before writing the letter. Moreover, letters from recent supporters will address any changes or new experiences that are critical to any future paths.
5. One must demonstrate success by whatever means that specific discipline recognizes as success. For example, in the scientific community, research is quite often the gauge by which success is measured, that is, grants, publications, presentations, etc. In this case, it must be recognized that collaborative and interdisciplinary research efforts are the "norm," that is, "team science" as larger and larger challenges in science are being undertaken. Appropriately, funding also follows a similar pattern as large awards are granted to collaborative teams. Plus, not only is this usually the most effective way to achieve success, but also especially relevant to the topic of this book; it leads to more diversity as diverse teams

are more creative, often due to the experiences that underrepresented minorities have with regard to adapting, overcoming, and coping. In general they are most flexible, excellent communicators and have the ability to take on multiple identities. As a final point to be made, in an area such as minority health disparities, the under represented student brings a personal exposure and commitment to the issue, which is always an advantage when addressing such major issues in research. At teaching institutions, evaluation of teaching performance and/or a role in curriculum development or reform may be a more critical assessment of success. In industry, it may be the development of a patent and/or a product with great profitability for the company. Whichever the case, one has to know what is viewed as success and then demonstrate success by those means to not only survive but to flourish. As a part of that, one has to have a sense as to what kinds of numbers are needed to meet the criteria. Some programs will actually list a number whereas other as more ambiguous; however, even in the latter case, insight can usually be gained relative to acceptable vs. unacceptable numbers. Just as an important note relative to publications and presentations, one has to be careful of "Red-flag conferences" (Brooks 2009 "Red-flag conferences") when building up their portfolios for tenure and promotion. As indicated in the article, these conferences have sprung up in recent years and are characterized by a number of things that should raise one's concerns from a professional perspective. For example, these are held at exciting destinations, for example, Las Vegas, have a presenter's fee just to attend, are advertised through a series of unsolicited emails, are low tech when it comes to presentations, have limited journals for publishing the works, as well as other "red-flags" that professionals should recognize. Participation in such conferences can be most damaging to one's career and in addition can cause problems at the institution level when accountability in attending such conferences is put to test. The bottom line is that faculty and other professionals have to recognize that there are such scams in academia and as such must avoid them at all costs, the biggest one being their professional integrity and credibility.

Minority Faculty Dissatisfactions

As directly relevant to the aforementioned last point, recent surveys have shown a higher degree of dissatisfaction among minority faculty members than their white counterparts, especially on their progress to tenure (Cleaver 2008, "Sense of fit can make or break faculty retention," *Diverse Issues in Higher Education*; Cropsey et al. 2008, "Why do faculty leave? Reasons for attrition of women and minority faculty from a medical school; 4 year results"; Jaschik 2008, "Racial gaps in faculty job satisfaction," *Inside Higher Education*; June 2008, "On the road to tenure, minority professors report frustrations," *The Chronicle of Higher Education* 12/5/08; Schmidt 2008 "Many Black women veer off path to tenure, researchers say," *The Chronicle of Higher Education* 9/9/08; Watson 2008 "Faculty and students of color face various dilemmas," *Diverse Issues in Higher Education; The Academy Speaks* 12/15/08; Gallagher and Trower 2009 "The demand for diversity," *The Chronicle of Higher Education*). Reasons include a lack of clarity about the tenure process and criteria, unfair and unequal treatment but, probably just as important, issues with the workplace culture and climate. The last issue relates to the sheer lack of numbers of "people who not only do not look like them but also do not relate to them," that is, an alienation, an expectation of extensive committee involvement, especially those committees related to diversity, their shouldering the burden of, and representing, all minorities, intolerance and confrontation from students because they "are different" and/or because they are presumed to be beneficiaries of affirmative action, and finally, the fact that they are in many ways prevented from advocating for their issues in that they would be "playing the race card." Plus very important to the last point, expending time and efforts in that arena will normally be most costly to their career since very few institutions or organizations use that as a criterion for advancement. In *The Chronicle of Higher Education* article by Gallagher and Trower, they provide a number of general strategies for recruiting and developing a diverse faculty. These include (1) visible leadership that is committed to this goal, (2) clearly defining a "diverse applicant pool" to search committees, (3) active recruitment efforts, (4) tapping into minority scholars, (5) establishing "target of opportunity" monies, (6) better educating search committees, (7) ensuring equal-opportunity mentoring, (8) showcasing successes, and (9) recognizing that culture is indeed personal. These are excellent strategies for accomplishing a diverse faculty that must be supported and implemented for such a change to occur. As a positive note, some institutions are indeed placing a value on efforts directed toward achieving diversity goals such as higher retention and graduation rates for minority students and as such are using some of these strategies. For example, the University of Virginia and the University of Maryland through commitments by the system's administrators as well as Kent State through incentives to faculty have demonstrated the types of efforts that are necessary and can profoundly affect the retention and graduation of underrepresented minority students (O'Rourke 2008, "Diversity and merit: How one university rewards faculty work that promotes equity"; Mahoney et al. 2008, "Minority faculty voices on diversity in academic medicine: Perspectives from one school"; Kinzie 2008, "At UVA, a dean making a difference"; Associated Press 2008a, "Kent State faculty get bonuses for meeting goals"; Fain 2007, "Sill Young enough to be hungry"; Associated Press 2008b, "Virginia Tech to increase investment in diversity"; Bass et al. 2008, "The university as mentor: Lessons learned from UMBC inclusiveness Initiatives"; Gose 2008, "Whatever happened to all those plans to hire more minority professors?"). Moreover, even though the points made earlier will help to ameliorate some of these issues, many major issues will still exist and in fact is the reason that the number of underrepresented faculty members in academia has changed very little in 30 years, for example, a 2007 US Department of Education report indicated that Blacks comprised 5.4% of the total faculty at all degree-granting institutions. In 1981, more than a quarter of a century earlier, that number was 4.2%! On the basis of this progress, it is estimated that it would take almost 150 years for the percentage of Black faculty in higher education to equal the percentage of blacks in the overall American workforce ("The snail-like progress of Blacks into faculty ranks," *Journal of Blacks in Higher*

Education 2009a). As discussed throughout this book, for the science areas, these numbers are even worse, especially at the nation's top research universities, for example, in 2007, underrepresented minorities made up only 3.9% of the chemistry faculty at the nation's top research universities. As a specific example, at Notre Dame, the percent Black faculty is at 2.3, which is the lowest of any of the nation's 25 highest-ranked universities ("Notre Dame takes steps to increase Black faculty," *The Journal of Blacks in Higher Education* 2009b). As a result of those numbers, Notre Dame has created a new initiative including a postdoctoral program to attract women and minorities and financial support to entice these scholars to stay in faculty positions.

General and Specific Tenure Issues

Since tenure is a critical aspect for all faculty members, and as noted in several references especially for minority faculty members, let me say a few words about it. First, a dictionary definition is "permanence of position," although in academia it was created essentially to "protect academic freedom and for all of the right reasons." For example, anyone who has been appointed to an administrative position is familiar with the phrase "You serve at the pleasure of…." In other words, there is essentially never any tenure with an administrative position; in fact, depending upon what level of an administrative position one is appointed to, there may not even be a contract. As a result, the individual can be removed for not doing their job, which most would consider "okay." However, there are many times when the person is removed, despite doing their job well but rather for "not pleasuring their supervisor" in whatever way that person wants pleasure. An example is when the person is not "conforming or fitting in," even though it may be for the right reasons and within the job description. On the other hand, tenured faculty members know that their positions are secure unless they commit a really heinous crime or act. For that reason, it is imperative that the individual knows what is required in their department, school, and university to attain tenure. As cited already, some minority faculty members indicate that lack of a clear definition of tenure was one of their main complaints within the system. A recent article in *Inside Higher Ed* (3/27/09), under the section "career advice," and written by Leslie Phinney (2009) addresses some concerns about tenure ("What I wish I'd known about tenure"). She lists nine facets of tenure and the tenure process (which need to always be discussed together!). These include the following: (1) The house sets the rules since tenure is a long-time investment by the institution and as such is dependent upon recourses and needs of the university, certainly not always faculty achievements; (2) it is like joining a fraternity and as such requires some adapting and co-opting, at least for awhile; (3) it is like a hunk of Swiss cheese in that there is substance/strength, that is, the cheese and weaknesses, that is, the holes, so build upon the cheese and eliminate or at least minimize the holes; (4) most of those embarking on the tenure journey will end up somewhere in the "gray zone" where decisions are not straightforward and easy; (5) like a disease, there a number of "risk factors" that should be recognized to be successful; (6) regardless of the "permanence of position," true tenure is being able to secure another position, so networking is critical; (7) preferably tenure matches one's ideals and values as permanence of position manes much less if one is not matched in an area where they are happy; (8) be ready for "fight or flight" throughout the process, and (9) as important as it is tenure is only one facet of life, even within academia, so treat it as such. These are just one person's compilation of thoughts, experiences, and ideas; however, as one who has been through the process at a top-ranked research institution, personally, I would agree and support these as being facets/issues that one wants to be aware of throughout their quest for tenure. Considering these, it is not too surprising that minority faculty as concerned about definitions about the process and the final decisions based on that process, since in many cases they are already not included in the "system" in the same way that nonminorities are. An excellent article by Metzler (2009, "The case against cultural standardization in tenure decisions") discusses some aspects about tenure decisions regarding Black faculty. In fact, he asks point blankly about the role that race plays in the tenure decision, especially in cases where tenure is denied. Not saying that all denials of tenure to Black faculty are cases of racism but rather he asks if the process, as exhibited by the committees, the departments, the deans, the provosts, and the presidents, has become "homogenized, structured around amorphous standards of scholarship and service, such that it is more likely than not that Black scholars and our scholarship will forever be relegated to the intellectual margins." Moreover, the question arises as to whether tenure decisions have become more about what is legally defensible than whether or not they are just. This represents a significant concern in light of the number of hires of minority faculty due to direct or perceived pressures to "diversify" their faculty. In addition, as the nature of teaching and leaning continues to change and evolve due to the students and as society becomes more multicultural and multiracial, the promotion and tenure process has to recognize, acknowledge, and eventually reward efforts directed toward this change. Heretofore, he states that this process has been a "bastion of pettiness, antagonism and ethnocentric backslapping." As such, his suggestion is that the rules for tenure and promotion resist cultural standardization and instead become specific, particular, and transparent. Alternatively Dr. Metzler suggests that tenure be abolished in favor of a system that rewards quality, inclusive scholarship and service. The latter would definitely address some of the concerns expressed by minority faculty about the tenure process.

Gender Differences and Considerations

It is important to note at this stage gender difference also become a problem but interestingly in a different way than alluded to earlier. In other words even though the numbers of minority women entering educational paths and actually completing them are more than with men, there appears to be a higher attrition rate especially in academia for minority women versus men. In fact, in a report commissioned by the National Association for Equal Opportunity in Higher Education, it was seen that Black women are quite likely to go from postdoctoral fellowships to actually being unemployed. (Schmidt 2008, "Many Black women veer off path to tenure, researchers say"; Lederman 2007, "Why women leave academic medicine"; Cropsey et al. 2008, "Why do faculty leave? Reasons for attrition of women and minority faculty from a medical school; 4 year results"; Watson 2008, "Faculty and students of color face various dilemmas"; Cleaver 2008, "Sense of fit can make or break faculty retention"; Branch-Briosos 2009, "Keeping pace, but not catching up"). An excellent discussion of these factors and how possibly to address them is provided in a published interview in *Inside Higher Education* (2/2/09) by Jaschik (2009) with Sandra Hanson (2008), the author of the book "Swimming against the tide. African American girls and science education." One of the most interesting results of the surveys that she reported was that the interest in science among young Black girls was high and in fact sometimes even higher than that for white girls, dispelling the rumor that Black girls just were not interested in science. Other issues that were identified include the lack of acceptance as Black women as scientists, the effects of institutional sexism and racism on their self-confidence, attitudes and achievement, inadequate school resources, the reliance on standardized test scores for admissions, negative stereotyping, the lack of role models and peers, and cultural/ family issues, among others. Of course, there are other issues that lead to attrition of minority women in academia, some of which are not related to ethnicity, for example, the "childbearing aspect" that requires women to take medical leaves at some point during their pregnancy regardless of "child rearing concerns" later. Unfortunately, in the fast-paced scientific arena, any time away from the normal routine of research/teaching can have devastating effects on one's career. In addition, even though the number of women entering the field is increasing, males still dominate the ranks especially at the decision-making positions and, as such, the environment, which is critical to success, can be less than inviting and/or favorable, in fact, in some cases hostile or worse. For example, when I was a faculty member at the University of Michigan Medical School and working with minorities at the student, staff, and faculty levels, a cultural audit was done by an external firm. Their findings included a statement that for Blacks in the medical school, "the environment at times was toxigenic." One does not have to be a toxicologist to recognize the severity of that environmental description! Interestingly, when I filed a lawsuit regarding the unfavorable climate for minorities at the medical school, the Public Relations department indicated that there was no basis for my claim. Another example was demonstrated in the case of Donna Shalala (Branch-Briosos 2009, "Keeping pace, but not catching up") when she was told by the chair of the department at her first teaching job that "her stellar performance didn't really matter. We have never tenured a woman and never will; it's a bad investment." Obviously, on the basis of Dr. Shalala's successes (she is now President of the University of Miami but additionally has been President of Hunter College, Chancellor at the University of Wisconsin and Health and Human Services Secretary under President Clinton), this attitude did not hold her back but this is not always the case; in fact, it is probably the exception rather than the rule. According to a report by the Association of American Colleges and Universities, women comprise 57% of the undergraduates at US colleges and earn the majority of the doctorates awarded to US citizens. In fact, according to the 2007 Survey of earned Doctorates, 51.4, 58.7, and 67.4% of the doctorates in biology, social sciences, and education were awarded to women, respectively. However, an American Association of University Professors report also showed that women were being hired at higher numbers into nontenure-track positions where their chances for promotions and salary increases were limited and that women held just 31% of tenured faculty posts and 45% of the tenure-track positions. A specific example is that of psychology where 72.6% of the doctorates are awarded to women but only 53.9% of the faculty in psychology are women and of that number only 42% are tenured. The reasons for these low numbers are many including those alluded to earlier such as the uncomfortable environment, search committees comprising mainly men, and lack of institutional commitment. In fact, in the Branch-Brioso article, those women quoted in the article, including Dr. Shalala, indicate that this problem must be attacked from all directions, that is, not only a commitment from the top, for example, President/Provost but also from the bottom–up, e.g., departments and faculty. As examples, UC Davis Professor of Law Emerita Martha S. West enlisted the assistance of a state senator whereas the University of Missouri at St. Louis uses its Office of Equal Opportunity to shut down searches if the search process is not demonstrating the type of diversity that would ensure a diverse pool. Furthermore, as the assignments become more prestigious, whether it is a tenured position at a doctoral granting institution or a position such as President, the number of women appointed to these positions are less, for example, 26% of tenured positions at doctoral-granting universities are women and only 23% of college and university Presidents are women, falling to just 14% at doctoral universities. In support of these numbers, a report from the 2009 American Council on Education conference not only showed that women and minorities are underrepresented among chief academic officers but also that they have a difficult time rising to the post of President (Schmidt 2009, "Survey of chief academic officers raises concerns about diversity and longevity"). Specific data showed that 85% of the chief academic officers were white, 6% Black, 4% Hispanic, 2% Asian American, and 1% Native American and furthermore, those from minority groups were found disproportionately at minority-serving institutions. On a positive note, with regards to such appointments, that is, minority women, Muriel A. Howard was recently appointed as President of the American Association of State Colleges and Universities (AASCU), the first time that a woman from a racial minority group will head one of the six major "presidential" associations (Lederman 2009, "New chief of state college group"). In this position, Ms. Howard, previously President of State University of New York (SUNY) at Buffalo, oversees the 430 4-year public institutions that comprise the AASCU. She states, as a Black woman, "I just hope we are the last generation of firsts, but I hope there are many more such firsts," adding "if I am a role model for other young women, for children of color, I'm ready." The timing of an event such as this is not coincidental, that is, with the "age of Obama," which was discussed recently and reported in an article by Nealy (2009b) in *Diverse Issues in Higher Education* online (4/1/09) entitled "Scholars discuss Black power in the age of Obama." Interestingly, as the Black Power movement is looked at historically, both positively and negatively, the picture of a multifaceted movement orchestrated by a number of constituencies of Black America is not often painted for what it contributed, which was discussed in depth during various workshops examining the role of women scholars and activists during the Black Power movement. Dr. Johnnetta Cole, director of the Smithsonian Institution's National Museum of African Art and former president of Bennett and Spelman Colleges spoke of those early days when there were essentially two, her and Ester M.A. Terry, associate chancellor and professor/chair of Afro-American Studies Department at Amherst, and how this absence of Black female scholars and their scholarship diminished significantly the impact of the discipline. She referred to the book titled *All of the Blacks are Men, All of the Women Are White, But Some of Us Are Brave* as having a profound effect on her realization of the relationship between her scholarship and her personal identities. At the same time, Black feminists such as Toni Cade Bambara, Alice Walker, Nikki Giovanni, and Toni Morrison demonstrated strong voices through their writings. And, not to be diminished at all during this Black Power time are the actions of the students such as those at Northwestern University who occupied the university bursar's office. From all of these actions, the statement made by these scholars from this symposium on the impact of the Black Power movement in America (in this Nealy article) is that "The Black Power movement is not a vestige of the past but a living didactical legacy that is as relevant now in the Obama era as it ever has been," as evidenced by the rise of not only those mentioned minority women scholars to various positions of leadership, such as Dr. Cole but others including Julianne Malveaux, a long-time activist for civil rights and now President of Bennett College for Women, and Phoebe A. Haddon, recently selected as the first Black to serve as dean of the University of Maryland Law School (Roach 2009, "Haddon selected as University of Maryland Law School dean, marking first for an African–American"). And, not to be forgotten as an integral part of how this recognition of minorities and particularly minority women has evolved to where it is now is the stance taken by Professor Derrick Bell when he was a law professor at Harvard to give up his position because of the lack of minorities and women on the Harvard law faculty. It is those types of commitments that are necessary for changes to occur, for example, the appointment of Dr. Wendell Pritchett as the first Black Chancellor for the Rutgers campus (Staff 2009, "Rutgers names first Black chancellor"). Still, despite these significant accomplishments and appointments of Black women such as those mentioned, an article by Anyaso (2008, "Self navigating THE terrain") discusses the fact that the absence of mentorship for Black women scholars could ultimately result in a dwindling pipeline as many of these individuals are left to navigate the personal and professional politics of the academy on their own. This is never a good thing especially when the numbers are insufficient to begin with. An additional factor in the lower numbers seen for women faculty is the fact that pay differences between men and women still exist and, as such, contribute negatively to the retention of women in the academic ranks, especially for minority women since a disparity still exists in pay equity between minorities and nonminorities. Still, a major contributor to the attrition numbers, particularly among underrepresented minority female faculty in the sciences, is the sense of "not fitting," due to the low numbers, which in turns leads to a "dwindling pipeline." This is directly related to the mentoring, or lack thereof, which is so critical, especially in the high-pressure world of academia and the "tenure process." Unfortunately, this dwindling pipeline effect will directly offset, in time, the increased numbers seen in recent years. So, it is critically important for minority and women academicians to be as visible as possible, that is, constant networking. In many cases the onus will be on the individuals to optimize the situation by placing themselves in the right environment. Venues for maintaining networking opportunities include professional societies and groups, for example, MentorNet, local and national conferences, for example, Compact for Faculty Diversity's Institute on Teaching and Mentoring, and just being sure to interact with individuals in any and all professional and personal settings, as one only realizes what "a small world it is" when they experience this for themselves, for example, being recognized in an airport, a conference, or the mall. Everyone can be a resource. Other programs that have been designated to address the gender issue in science include the ADVANCE program from NSF (Adams 2008, "Nurturing women scientists"). The University of Wisconsin was one of those awardees and approached the problem by educating the faculty, particularly those who served on hiring and tenure committees, on research-based evidence on unconscious bias. Another awardee was Rensselaer Polytechnic Institute (RPI), which created a program called RAMP-UP (Reforming Advancement Processes Through University Professions). RPI President Shirley Ann Jackson, who has been a most successful minority woman administrator, focused that program on both progression of the women faculty through the ranks and also on expanded recruitment of women at the senior rank. Key for success with all of these programs is mentoring, especially relative to changes in culture at the institutions.

In the case of women in STEM, the work-related discrimination, of course, crosses all ethnic lines since the males dominated the STEM field. Chubin and Sevo (2008) discuss this as part of the Project Working WISE in a paper entitled "Work-related discrimination in STEM workplaces." Starting with a discussion of "what is a fair workplace?" the authors then address the underlying principles behind the discriminatory actions, which, as it turns out, are often just habits and preferences rather than policies and procedures. So then what are the obstacles to change? The authors identify at least four major ones, including ignorance, conceptions of what constitutes discrimination, the inequalities that exist in home and family that carry over to the workplace, and the ineffectiveness of current laws, policies, and models designed to prevent discrimination. Identifying these constitutes an important step in addressing discriminatory practices whether they be gender based or based on ethnicity, sexual preference, etc.

Institutional Racism

At this point it is important to discuss the effect of institutional racism, both from the past, in the present, and of course in the future. Much of the underrepresentation of ethnic minorities has been due in some part to the effects of institutional racism.

And to be clear, institutional racism does not refer only to racism at the level of the academic institution itself or even in the academy in general but rather to that exhibited as part of the society in which we live. This institutional racism is tied directly to the white privilege issue that has been discussed previously, resulting in discrimination, biases, and inequalities, which in turn result in underrepresentation in a number of areas. In fact, Dr. Lovell Jones (2009) asks a very important question in his article entitled "Don't go along to get along," which appeared in 2009 in *Progress Magazine*. That question was "has anyone done a longitudinal study on the impact of workplace racism stress and health?" He states that this stress is what people of color face who "don't go along to get along" and that is the rule rather than the exception! He further goes on to say that ultimately it is either a case of "stress of confrontation" or "stress of assimilation," resulting in losing no matter what. His argument is only fortified by looking at the differences between blacks and whites with regard to such things as unemployment levels (9.0% vs. 4.2%), median household income ($35,464 vs. $63,156), home ownership (48.2% vs. 75.8%), college graduation rate (41% vs. 61%), life expectancy (73.3 years vs. 78.3 years), etc. And, with the most recent economic situation, and the rising unemployment, the groups that are hardest hit are those individuals of color, that is, Latinos and Blacks (Associated Press 2009, "Unemployment hits harder among Latinos, Blacks"). Statistically, since December 2007, Latino unemployment rose 4.7% points to 10.9% whereas during the same time, Black unemployment rose 4.5 points to 13.5%. White unemployment rose 2.9 points to 7.3%, once again approximately almost 50% of what the numbers are for Blacks. Obviously, the less wealthy are the harder hit with unemployment and considering the 2002 data that showed Black households with a median net worth of $6,000, Latinos with $8,000 and whites with $90,000, there is no question of the impact of this rise in unemployment. In fact, in that *Diverse Issues* article, Duke Professor William Darity from the Economics and African–American studies departments states "America has never come to terms with racial economic inequality and the current situation is reinforcing and widening those inequalities." Moreover, when looking at all of these figures, one fully realizes why underrepresentation, particularly in STEM where it has traditionally been predominately white and male, has been, is, and will continue to be a problem, not just in academia but across our society. Unfortunately, changes in differences of this magnitude require a complex set of actions, most likely to occur over the course of a significant amount of time. One obvious change that has to occur is that there must be more open discussions about race and racism. Paul Lyons, a faculty member at Richard Stockton College in New Jersey, before his death in January 2009, wrote in his teaching log "I don't think we as academics and teachers do a very good job teaching about race and racism. Some sees to be liberal guilt. Mostly it rests on the lack of confidence that one can present complicated situations, nuanced realities without risking being misinterpreted by colleagues and students." Lyons goes on to discuss other situations, including involvement with panels that exemplify the complexity of such discussions but also how important it is to have these discussions openly and honestly. As almost any faculty member how they deal with race issues in their classroom, especially in the science area and most will say "I don't." Another crucial example, as mentioned early on, the specific funding of programs to address this underrepresentation, has been in place for almost one half of a century, albeit at levels much too low for what is needed, with gains that at times seem almost negligible. Thus, the efforts have not equaled the need, most likely due to the lack of prioritization of the problem. An excellent example is the issue of minority health disparities, which has developed into one of the most significant problems facing our society today (in fact, in an article referenced earlier, this issue was called "the most prevalent civil rights issue of the decade" Nealy 3/16/09) and will only continue to worsen as a problem as our demographics continue to change. Under President Clinton, the then Surgeon General, with the support of that administration, identified this issue as a priority and, in fact, set a goal to eliminate this as a problem by 2010. Efforts were definitely put forward to address this issue; however, with the change of administration, came a significant change in priorities, resulting in health disparities no longer being "as important and therefore not as high of a priority." Thus, this was a case of the problem not becoming less of an issue; in fact, it became even more of an issue as the minority population continued to grow, but rather, what happened was the problem became less of a priority due to those in power to make the decisions not considering it as high of a priority to address, due in large part to white privilege and the resultant institutional racism. Moreover, it has to be recognized that the issue of health disparities is truly one that involves much more than medicine and personal skills of the healthcare providers, that is, the role that political, social, economic, and cultural factors play cannot be underestimated. Obviously, this relates very much to an issue that was also mentioned previously and that is the sheer lack of numbers of minorities and/or nonminorities who are sensitive to, and support of, minorities, who are in decision-making positions of power. So, although Clinton's most favorable support of minority issues coupled with Dr. Satcher's appointment was a very critical step in effectively addressing this issue, those efforts diminished significantly when the change occurred and, in fact, actually set back some of the progress that had been observed. Certainly, when considering the election of Barack Obama in 2008 and his commitment to science, health, and diversity, coupled with his appointments to key persons in his administration, the hope is that by prioritizing issues such as minority health care and numbers of minorities in the workforce, just to name two, the issue of institutional racism can be addressed much more effectively nationally. There will not only be a prioritization on these issues by the powers to be but also an awareness of the importance of these issues nationally and globally that is also critical in getting things to happen regarding this issue, regardless of minority or nonminority status. The battle is still an uphill one as even though the numbers of minorities are increasing, insufficient numbers of them are in decision-making positions. Plus, since the privilege of being white is a major contributor to institutional racism, more whites have to be willing to give up some of that privilege to effectively address the racism (Wise 2005, "White like me: Reflections on race from a privileged son"; Wise 2009, "Between Barack and a hard place: Racism and white denial in the age of Obama"; Thompson et al. 2004, "White men challenging racism"). Realistically, first of all how many people are willing to give up, that is, sacrifice a privilege and second, give that up to address an issue that is a problem but not necessarily one that affects them directly? In fact, Dr. Achebe in the Nealy (2009a) article (*Diverse issues* 3/16/09) notes that the physicians do not have to be Black, Hispanic, or Asian to address disparities and cites the fact that most of those working in his community health center are nonminorities and that the Chief of Gynecology is one of the most culturally competent doctors that he knows. However, without this happening more globally, the institutional racism will continue to exist and as a result, especially in fields such as STEM, those who are not traditionally part "of the group" will be disadvantaged at all levels resulting in continued underrepresented numbers, insufficient to address this disparities issue.

References

Adams JU (2008) Nurturing women scientists: Nationwide and institution-sized surveys show a leaky pipeline partially patched, but the reservoir still far from full. Science http://sciencecareers.sciencemag.org/career_magazine/articles/2008_02_08

Anyaso HH (2008) Self navigating THE Terrain. Diverse Issues in Higher Education online http://www.diverseeducation.com/artman/publish/2008/11/13

Associated Press (2008a) Kent State faculty get bonuses for meeting goals. Diverse Issues in Higher Education online http://www.diverseeducation.com/artman/publish/2008/09/11

Associated Press (2008b) Virginia Tech to increase investment in diversity. Diverse Issues in Higher Education online http://www.diverseeducation.com/artman/publish/2008/09/02

Associated Press (2009) Diverse issues in Higher Education online http://www.diverseeducation.com/artman/publish/2009/03/24

Bass et al (2008) The university as mentor: Lessons learned from UMBC inclusiveness Initiatives. Council of Graduate Schools: Volume one in the CGS Occasional paper series on Inclusiveness

Branch-Briosos K (2009) Keeping pace, but not catching up. Diverse Issues in Higher Education online http://www.diverseeducation.com/artman/publish/2009/03/05

Brooks M (2009) Red-flag conferences. The Chronicle of Higher Education http://chronicle.com/jobs/news/2009/03/2009030260lc Accessed 3/26/09

Chubin D, Sevo R (2008) Work-related discrimination in the STEM workplace. Working WISE : Intergenerational voices advancing research and policy for women in science, technology, engineering and mathematics. In: Rayman P, Bond MA, Brunette M, Lally J, Hunt E (eds)

Cleaver S (2008) Sense of fit can make or break faculty retention. Diverse Issues in Higher Education online http://www.diverseeducation.com/artman/publish/2008/12/11

Cropsey KL, Masho SW, Shiang R, Sikka V, Kornstein SG, Hampton CL (2008) Why do faculty leave? Reasons for attrition of women and minority faculty from a medical school; four year results. J Woman's Health 17(7):1–8

Fain P (2007) Still Young enough to be hungry. The Chronicle of Higher Education. http://chronicle.com Accessed 4/11/07

Gallagher A, Trower CA (2009) The demand for diversity. The Chronicle of Higher Education http://chronicle.com/jobs/news/2009/02/2009020401c Accessed 2/4/09

Gose B (2008) Whatever happened to all of those plans to hire more minority professors? The Chronicle of Higher Education http://chronicle.com/weekly/v55/i05/05b Accessed 9/26/08

Hanson SL (2008) Swimming against the tide. African American Girls and Science Education. Temple University Press, Philadelphia

Journal of Blacks in Higher Education (2009a) The snail-like progress of Blacks into faculty ranks http://www.jbhe.com
Jaschik S (2008) Racial Gaps in Faculty Job Satisfaction. Inside Higher Education http://www.insidehighered.com/news/2008/12/05
Jaschik S (2009) Swimming against the Tide. Inside Higher Education http://www.insidehighered.com/news/2009/02/02
Jones LA (2009) Progress Magazine Don't Go Along to get along http://www.justgarciahill.org/jghdocs Accessed 1/9/09
Journal of Blacks in Higher Education (2009b) Notre Dame takes steps to increase Black faculty http://wwwjbhe.com/latest/index Accessed 3/26/09
June AW 2008 On the road to tenure, Minority Professors report frustrations. The Chronicle of Higher Education http://chronicle.com/daily/2008/12/8030n Accessed 12/5/08
Kinzie S (2008) At U-VA, a dean making a difference washingtonpost.com Accessed 5/18/08
Lederman D (2007) Why women leave academic medicine. Inside Higher Education http://www.insidehighered.com/news/2007/09/21 Accessed 9/2/08
Lederman D (2009) New chief for state college group. Inside Higher Ed http://www.insidehighered/news/2009/04/01
Lyons P (2009) Encouraging political incorrectness and civility. Inside Higher Ed http://www.insidehighered.com/news/2009/03/31
Mahoney MR, Wilson E, Odom KL, Flowers L, Adler SR (2008) Minority faculty voices on diversity in academic medicine: Perspectives from one school. Acad Med 83:781–786
Metzler CJ (2009) The case against cultural standardization in tenure decisions. Diverse Issues in Higher Education online The Academy Speaks http://diverseeducation.wordpress.com/2009/04/06
Nealy MJ (2009a) Racial health disparities called most prevalent civil rights issue of decade. Diverse Issues in Higher Education online http://www.diverseeducation.com/artman/publish/2009/03/16
Nealy MJ (2009b) Scholars discuss Black power in the age of Obama. Diverse Issues in Higher Education online http://www.divsereducation.com/artman/publish/2009/04/01
O'Rourke S (2008) Diversity and merit: How one university rewards faculty work that promotes equity. The Chronicle of Higher Education http://chronicle.com Accessed 9/26/08
Phinney LM (2009) What I wish I'd known about tenure. Inside Higher Ed http://www.insidehighered.com/advice/2009/03/27
Roach R (2009) Haddon selected as University of Maryland law School dean, marking first for African–American. Diverse Issues in Higher Education online http://www.diverseeduction.com/artman/publish/2009/04/06
Schmidt P (2008) Many Black women veer off path to tenure, researchers say. The Chronicle of Higher Education http://chronicle.com/news Accessed 9/9/08
Schmidt P (2009) Survey of chief academic officers raises concerns about diversity and longevity. The Chronicle of Higher Education http://chronicle.com/daily/2009/02/11231n Accessed 2/10/09
Staff (2009) Rutgers names first Black chancellor. Diverse issues in Higher Education online http://www.diverseeducation.com/artman/publish/2009/04/06
Thompson C, Schaefer E, Brod H (2004) White men challenging racism. Duke University Press, Durham
Watson E (2008) Faculty and students of color face various dilemmas. Diverse Issues in Higher Education: The Academy Speaks online http://www.diverseeducation.com/artman/publish/2008/12/15
Wise T (2005) White like me: Reflections on race from a privileged son. Soft Skull Press, New York
Wise T (2009) Between Barack and a hard place: Racism and white denial in the age of Obama. City Lights Books, San Francisco, CA

Chapter 9
Wrap Up and Guidelines

This book has included a number of recommendations and suggestions on how individuals can take full advantage of the opportunities that are available to them to achieve success at various stages of their careers, all the way from K-12 through college, their formal educational preparation, e.g., medical school, graduate school, and finally in their careers. At this point, it is important to simplify those steps that are needed so that a 'how to" list is readily available.

This chapter is designed to summarize the points discussed in the preceding chapters; in fact, it is designed as a "cheat sheet" for all those individuals who need mentoring in preparing for careers in science, professionals who need to stay on "top of their game" and generally anyone who just needs some tips on becoming successful and stay there.

1. Let us start with your *academic performance*
   - Always perform to the best of your ability
   - Continually assess your performance and standing in the class
   - Identify any problems as early as possible and seek help if needed
   - Identify resources as well as others that can help you
   - Use resources outside of the class to the maximum, e.g., library, Internet
   - Stay updated on all prerequisite information and requirements

2. The Preparation of the *personal statement* is very important and you should
   - Identify your top 2–3 qualities as the focus of your statement
   - Make the first sentence an "attention getter"
   - Take a historical approach as a start
   - Make first sentence of each paragraph memorable
   - Transition from one paragraph to another
   - Be unique
   - Speak to process, not just results
   - Be specific where appropriate
   - Make sure they are aware of your potential
   - Limit "I" references
   - It must be a story

T. Landefeld, *Mentoring and Diversity,* Mentoring in Academia and Industry 4,
DOI 10.1007/978-1-4419-0778-3_9, © Springer Science+Business Media, LLC 2009

- Be strong and confident, but not arrogant
- Provide explanations without apologies
- Informal is okay but avoid slang
- Definitely show your personality and your human side
- Highlight your interests that truly reflect you
- Be sure to make the statement say what you want it to say
- Do not step on any toes

3. In choosing a mentor and those who will write you a *letter of recommendation*
   - Identify potential mentors and score them with "+" and "–" using categories such as taking classes with them, performance in classes with them, number of times that you met with them outside of class, the type of rapport that you have with them, etc. They must be a "good fit"
   - Choose those with the highest marks (most "+"'s) for mentors
   - Meet with them to request that they be your mentor
   - Provide them with regular and frequent updates as to your progress
   - Follow up your meetings with them with an email
   - Inform them well in advance what your plans are and how they can help
   - If this involves writing a letter of recommendation, provide them with the information well in advance
   - Ask them if they will share their letter with you (very important)
   - Gently remind them in advance of when the letter is needed
   - Once the letters are completed, write a thank you note
   - Keep them updated as to your progress with the process
   - Maintain contact and update progress with them even after completion of the letter and in fact even after you enter the program of study

4. When preparing for an interview and actually interviewing
   - Present yourself professionally
   - Use a strong handshake
   - Learn all that you can about the program/school
   - Always maintain eye contact
   - If possible, learn about the interviewer
   - Have questions ready when asked "do you have any questions for us?"
   - Have at least one mock interview with someone familiar with the process
   - Be confident but not arrogant
   - Demonstrate enthusiasm but don't be "giddy"
   - Know your goals and skills well
   - Avoid negative comments about anyone
   - Answer as succinctly as possible
   - If you do not know the answer, say that you don't know
   - Repeating the question often helps to get your thoughts together

5. Your portfolio needs to be kept updated at all times as you never know when you will need it; in fact, when you do need it usually it is an immediate need and just like websites it never should be out of date. Although this should not be a

problem with word processing, outdated portfolios are usually just from neglect and lack of attention. Although the following list of recommendations may seem like "givens," it is important to note them anyway
   - Develop and maintain a resume and a curriculum vita
   - Be professional in all aspects of your personal presentation
   - Update changes in contact information, e.g., phone, fax, email, address
   - Include all recent publications, abstracts, awards, achievements
   - Keep references, both new and established updated with recent information
   - Include all activities that have relevance to your credentials/experience

6. Networking will continue to be one of the most important Aspects in your career development and advancement. As such you must always
   - Have business cards available
   - Have copies of your resume when attending any type of conference
   - Be proactive (but not too aggressive) in introducing yourself to others
   - Give a brief description of yourself & your goals when introducing yourself
   - Be professional in your appearance
   - Get a business card from the people that you meet
   - Follow up with the person after meeting them
   - Continue to stay in touch on a regular basis
   - Keep them updated as to your progress/accomplishments

7. Some general tips for accomplishing your goals
   - Know who you are, both your strengths and weaknesses
   - Always be yourself
   - Be passionate about something
   - Do not accept the norm as the only way
   - Do not be afraid of making mistakes
   - Learn from your mistakes
   - Choose people around you carefully
   - Always be aware of all opportunities and how to take advantage of them
   - Recognize that it will often be difficult ("Without struggle, there is no progress" – F. Douglass)
   - Always keep "your eyes on the prize"
   - Be confident but not arrogant
   - Always remain humble
   - HAVE FUN and ENJOY

Finally, just as a reminder of what a mentor and mentoring does, as you will always have a mentor and will always be a mentor:

- Teaches
- Provides support
- Provides advice and counseling
- Provides advocacy
- Provides networking opportunities

- Provides constructive criticism
- Shares in your goals and dreams
- Broadens your exposure
- Believes in you

No matter how busy you become, always make time to mentor, remembering that someone did that for you and without that mentoring, you would not be who or where you are now. Remember Dr. King's quote; "Everyone can be great because everyone can serve."

Index

Breinigsville, PA USA
18 October 2009

226007BV00004B/35/P

9 781441 907776